A Common Operating Picture for Air Force Materiel Sustainment

First Steps

T0308625

Raymond A. Pyles, Robert S. Tripp,

Kristin F. Lynch, Don Snyder,

Patrick Mills, John G. Drew

Prepared for Project Air Force

Approved for public release; distribution unlimited

PROJECT AIR FORCE

The research described in this monograph was sponsored by the United States Air Force under contract FA7014-06-C-0001. Further information may be obtained from the Strategic Planning Division, Directorate of Plans, Hq USAF.

Library of Congress Cataloging-in-Publication Data

A common operating picture for Air Force materiel sustainment : first steps /
    Raymond A. Pyles ... [et al.].
        p. cm.
    Includes bibliographical references.
    ISBN 978-0-8330-4128-9 (pbk. : alk. paper)
        1. United States. Air Force—Equipment and supplies. 2. Airplanes, Military—
United States—Maintenance and repair. 3. Logistics. I. Pyles, Raymond, 1941–

UG1103.A18 2008
358.4'180973—dc22

                                                        2008025824

The RAND Corporation is a nonprofit research organization providing objective analysis and effective solutions that address the challenges facing the public and private sectors around the world. RAND's publications do not necessarily reflect the opinions of its research clients and sponsors.

RAND® is a registered trademark.

Published 2008 by the RAND Corporation
1776 Main Street, P.O. Box 2138, Santa Monica, CA 90407-2138
1200 South Hayes Street, Arlington, VA 22202-5050
4570 Fifth Avenue, Suite 600, Pittsburgh, PA 15213-2665
RAND URL: http://www.rand.org/
To order RAND documents or to obtain additional information, contact
Distribution Services: Telephone: (310) 451-7002;
Fax: (310) 451-6915; Email: order@rand.org

Preface

This monograph outlines how the U.S. Air Force could develop and use a common operating picture to plan and execute materiel sustainment activities worldwide. First and foremost, the monograph describes a methodology for developing operationally relevant, effects-based metrics that can be used to plan and control materiel sustainment processes, while taking into account the unique local observations and diverse knowledge base provided by widely dispersed sustainment infrastructure and personnel. Then it applies that methodology to the problem of planning and managing the sustainment of depot-level reparable (DLR) components and the Air Force's Expeditionary Logistics for the 21st Century (eLog21) initiatives being implemented to transform the Air Force materiel sustainment system. The monograph pays special attention to how the Air Force Materiel Command's (AFMC's) new Global Logistics Support Center (GLSC) might exploit a common operating picture based on effects-based measures.

Thus, we envision three audiences for this monograph: policymakers interested in furthering the Air Force's eLog21 transformation, analysts and managers charged with guiding and implementing that transformation, and the specific managers charged with developing the GLSC processes and information architecture. While this document focuses sharply on the relatively narrow issue of DLR support, we believe that future analyses using this methodology could substantially improve the planning and management of all Air Force sustainment activities. We also believe that the GLSC initiative now under way could transform the worldwide logistics support processes to move beyond remedying DLR imbalances by lateral resupply to minimizing these imbalances by controlling the DLR sustainment system's resources and processes.

The research described in this monograph was conducted within the Resource Management Program of RAND Project AIR FORCE and was prepared as part of the fiscal year 2006 project "Implementing the Air Force Process Efficiency Initiative." The work was sponsored by the Deputy Chief of Staff for Logistics, Installations, and Mission Support (AF/A4/7). This report is one of a series of RAND reports that address planning and execution functions in support of agile combat support (ACS) in the Air Force. For other perspectives on the planning and management issues involved in the operation of that large, complex, and diverse enterprise, readers may find useful insights in the following titles:

- *Supporting Expeditionary Aerospace Forces: An Integrated Strategic Agile Combat Support Planning Framework*, Robert S. Tripp (MR-1056-AF), 1999. This report describes an integrated combat support planning framework that may be used to evaluate support options on a continuing basis, particularly as technology, force structure, and threats change.
- *Supporting Expeditionary Aerospace Forces: An Operational Architecture for Combat Support Execution Planning and Control*, James Leftwich, Robert S. Tripp, Amanda B. Geller, Patrick Mills, Tom LaTourrette, Charles Robert Roll, Jr., Cauley Von Hoffman, and David Johansen (MR-1536-AF), 2003. This report outlines the framework for evaluating options for combat support execution planning and control. The analysis describes the combat support command and control operational architecture as it is now and as it should be in the future. It also describes the changes that must take place to achieve that future state.
- *Supporting Air and Space Expeditionary Forces: Expanded Operational Architecture for Combat Support Planning and Execution Control*, Patrick Mills, Ken Evers, Donna Kinlin, and Robert S. Tripp (MG-316-AF), 2006. This report expands and provides more detail on several organizational nodes in our earlier work

that outlined concepts for an operational architecture for guiding the development of Air Force combat support execution planning and control needed to enable rapid deployment and employment of Air and Space Expeditionary Force (AEF). These combat support execution planning and control processes are sometimes referred to as Combat Support Command and Control processes.

- *Strategic Analysis of Air National Guard Combat Support and Reachback Functions*, Robert S. Tripp, Kristin F. Lynch, Ronald G. McGarvey, Don Snyder, Raymond A. Pyles, William A. Williams, and Charles Robert Roll, Jr. (MG-375-AF), 2006. This report analyzes transformational options for better meeting combat support mission needs for the air and space expeditionary force (AEF). The role the Air National Guard may play in these transformational options is evaluated in terms of providing effective and efficient approaches in achieving the desired operational effects. Four Air Force mission areas are evaluated: centralized intermediate repair facilities within the continental United States, civil engineering deployment and sustainment capabilities, GUARDIAN[1] capabilities, and Air and Space Operations Center reachback missions.
- *A Framework for Enhancing Airlift Planning and Execution Capabilities Within the Joint Expeditionary Movement System*, Robert S. Tripp, Kristin F. Lynch, Charles Robert Roll, Jr., John G. Drew, and Patrick Mills (MG-377-AF), 2006. This report examines options for improving the effectiveness and efficiency of intra-theater airlift operations within the military joint end-to-end multimodal movement system. Using the strategies-to-tasks framework, this report identifies shortfalls and suggests, describes, and evaluates options for implementing improvements in current processes, doctrine, organizations, training, and systems.
- *The Closed-Loop Planning System for Weapon System Readiness*, Richard Hillestad, Robert Kerchner, Louis W. Miller, Adam C. Resnick, and Hyman L. Shulman (MG-434-AF), 2006. This

---

[1] GUARDIAN is an Air National Guard information system used to track and control execution of plans and operations such as funding and performance data.

report addresses the problem of planning balanced component-repair capacity and funding across multiple fleets in light of unpredictable demands for depot-level repair of multiple components by developing a "closed-loop" planning methodology that estimates the effect of depot repair funding allocations on aircraft availability.

RAND Project AIR FORCE

RAND Project AIR FORCE (PAF), a division of the RAND Corporation, is the U.S. Air Force's federally funded research and development center for studies and analyses. PAF provides the Air Force with independent analyses of policy alternatives affecting the development, employment, combat readiness, and support of current and future aerospace forces. Research is conducted in four programs: Aerospace Force Development; Manpower, Personnel, and Training; Resource Management; and Strategy and Doctrine. Additional information about PAF is available on our Web site: http://www.rand.org/paf/

Contents

Figures

Tables

Summary

The Air Force materiel sustainment system (MSS) is continually caught between two countervailing forces: demands for increased efficiency and lower costs on one side, and demands for increasingly effective support to combat operations and peacetime training on the other.[2] Compounding the situation, the Air Force is currently facing more unpredictable operational demands—in terms of both their location and their required operating capabilities.

We envision a materiel sustainment system common operating picture (COP) that would better synchronize the MSS's activities— enhancing responsiveness to changing operational needs, reducing opportunities for unintended wasted effort, and coordinating efforts to improve support in one agency while ensuring that complementary efforts in another area are accomplished.

Underlying Principles for a Common Operating Picture

The notion of a COP is one that resonates with many people who work in large organizations. Ideally, it would provide common guidance and progress assessment that all parts of the organization could use to

---

[2] We chose the term *sustainment* rather than *operations and support* because the former includes all the activities associated with operating a fleet throughout its entire life cycle, including modifying the fleet to increase its airworthiness (flight safety), efficiency, or effectiveness in its current missions or its ability to conduct other missions. Under this definition, we include activities that change the fleet's capabilities, not just those that restore them.

assess their own contributions and view their mutual achievements as an integrated whole. While it would set explicit goals for both overall system performance and individual agencies, it would also define sentinels—active monitors of leading indicators that signal when the overall system's performance may fall below intended levels.

To develop a COP that achieves the goal of synchronizing a large organization's activities, we turn to four organizing principles: effects-based measures, *schwerpunkt*, decision rights, and a nonmarket economic framework. Use of effects-based measures consists of first defining the goals or objectives that an organization wants to achieve, then defining (generally long-term) measures of effectiveness (MOEs) for associated goals, and, finally, deriving measures of effectiveness indicators (MOEIs) to monitor progress toward those goals. *Schwerpunkt*, which means "focal point," is a notion drawn from physics and German military doctrine that emphasizes achieving the commander's intent regardless of unplanned events that may render the original plan obsolete, with even the lowest-level subordinates given wide latitude to take independent action to respond more quickly to those unplanned events. Decision-rights theory provides a framework that helps one decide which agency or person should make a particular decision, based on the information, decision capacity, and incentives that each agency faces. Finally, the nonmarket economic framework recognizes that inefficient conflicts can arise when a competitive market does not exist and that, therefore, there is a need for a neutral integrator to help resolve conflicts between demand-side and supply-side agencies within large complex organizations such as the Air Force.[3] (See pp. 5–18.)

---

[3] In a military context, one can view the demand side as being operational forces that require personnel, equipment, materiels, fuel, and services; the supply side would be the combat support activities that provide those resources. Thus, the combat support activities would be the rough equivalent of the supply chain that stretches from the local store to a manufacturer in a commercial context.

A Procedure for Developing a Common Operating Picture

We then define a process for applying these principles. The process consists of eight steps:

1. Identify the organization's broader objectives.
2. Relate those objectives to effects that the supply side must produce.
3. Identify measures of effectiveness.
4. Identify the processes and decisions that can affect each MOE.
5. Identify practical, comprehensive MOEIs and alarm thresholds for each MOE.
6. Assign the decision rights for the demand side, supply side, and neutral integrator to the lowest-echelon agency with the appropriate information and decisionmaking capacity.
7. Adjust the incentives for that agency to reward performance against the relevant MOEIs.
8. Periodically review these steps and adjust the COP and decision rights accordingly. (See pp. 19–31.)

An Example: Depot-Level Reparables for Aircraft

We then applied that process to the activities associated with aircraft DLR sustainment, noting that other sustainment activities might contribute to other Air Force objectives. We identified five objectives associated with aircraft sustainment:

- operational suitability
- mission reliability
- airworthiness
- availability
- sustainment cost.

Then we narrowed in on the availability objective to define an MOE—aircraft not fully mission capable for DLRs (NFMCD)—

that is associated with DLR sustainment.[4] With that in mind, we then identified several MOEIs (aircraft operational tempo [optempo], item demand rates, and various pipeline quantities and trends) that might act as leading indicators. (See pp. 33–48.)

With the MOEs and MOEIs in hand, we then assessed the current DLR sustainment system's ability to plan, monitor, and control NFMCD levels consistent with both the available funds and the needs of the demand side. To perform that task, we identified the range of decisions made throughout the Air Force that can affect NFMCD and selected two key decisions for further analysis: planning financial resources for DLR sustainment and reallocating financial and other resources to meet changing operational requirements. Then we applied a decision-rights matrix to evaluate the information available to the many agencies whose decisions can affect NFMCD during planning and execution. Because the planning system does not use NFMCD consistently throughout the financial planning process, we evaluated how such a system might work if NFMCD were available. This led to a suggested organizational arrangement for DLR sustainment planning. As shown in Figure S.1, this organizational structure envisions that a newly created AFMC Centralized Asset Management office (AFMC CAM)[5] would act as a neutral integrator to help the demand side (the major commands [MAJCOMs] and the Air Force forces [AFFOR] assigned to combat commanders [COCOMs]) and the supply side (maintenance, supply, and transportation supply-chain agencies and suppliers whose efforts will be coordinated by the newly designated GLSC develop a long-term DLR support plan. The neutral integrator's

---

[4] This measure expands the scope of a related measure—not mission capable (because of) supply shortage (NMCS)—to encompass the full effects of the end-to-end DLR support enterprise. First, it includes situations in which local base repair shops are repairing a component that could make an aircraft fully mission capable (FMC). Second, it includes situations in which an unserviceable component leaves an aircraft partially mission capable (PMC), that is, able to perform only a subset of the full range of missions for which it was designed.

[5] CAM is both a "system" and an AFMC office within that system. The office plays a lynchpin role in planning and managing the allocation of funds to Air Force sustainment activities, thereby assuring responsive support to all the MAJCOMs and COCOMs—within the limits of available resources.

Figure S.1
**The AFMC Centralized Asset Management Office Is the Neutral Integrator
for DLR Sustainment Financial Planning**

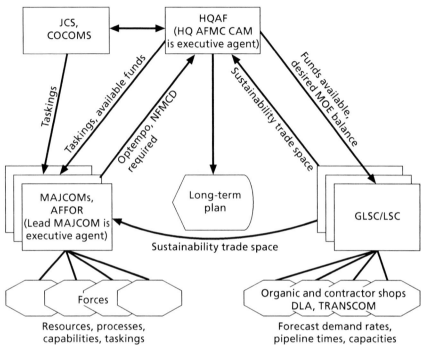

NOTE: The "GLSC/LSC" in the figure represents the Air Force's current transitional
situation, in which implementation of the GLSC is still under way and, therefore,
separate MAJCOM logistics support centers (LSCs) supplement the GLSC.
RAND MG667-S.1

role would be to ensure that the plan balances future support among the
various demand-side agencies based on operational needs and priorities
while also exploiting the supply side's financial and physical production resources to the fullest extent possible. To achieve this, the neutral
integrator (in this case, AFMC CAM) would seek explicit agreement
between the supply side and the demand side about the MOEs (operational tempo and NFMCD levels) to be achieved. (See pp. 48–98.)

In execution, the current DLR sustainment system has several
processes that can remedy the inevitable imbalances across fleets and
units when demands do not arise as predicted. Unfortunately, the cur-

rent processes can only mitigate imbalances once they become apparent; they cannot prevent imbalances from occurring. Because the DLR support system requires some repair and transportation times to respond to any imbalance, it must always play "catch up" if it relies solely on the current state for guidance.

To give the DLR sustainment system a more proactive posture, we identified how the new GLSC might monitor the NFMCD-based MOEIs during execution to detect underlying demand and DLR support process changes that might threaten the planned NFMCD level—before changes in that level reach unacceptable bounds. Thus, we suggest an organizational arrangement in which the GLSC acts as a central communications conduit and overall system monitor between the suppliers and the demanders of DLR sustainment activities. As shown in Figure S.2, the GLSC would monitor MOEIs (e.g., optempos and pipeline quantities) that reflect ongoing operations of all agencies in the supply side (including AFMC-owned and -operated agencies, such as the Depot Maintenance Activity Group [DMAG], and external suppliers, such as the Defense Logistics Agency [DLA], the Transportation Command [TRANSCOM], and the numerous contractors that repair DLRs) and the demand side (MAJCOMs and COCOMs). To support that process, the execution COP would embody active monitors of the MOEIs that we call sentinels. Those sentinels would periodically compare the MOEIs to their alarm thresholds, which would be derived from the NFMCD-based sustainment assessments developed for the planning COP. Both the MOEIs and the sentinels alarms would be available to the demand side and the supply side to monitor and control their own operations, while the GLSC would focus on those MOEIs that may signal some long-term, widespread, or other critical supply/demand side imbalance.

Almost certainly, some portions of the supply side will be unable to fully implement the detailed plan developed during the planning process. Equally probable, some portions of the demand side will find that their operational requirements change in response to new taskings or missions. In many cases in which the disruption to the original plan is short or small, the supply-side resources may be flexible enough to respond to those changes without breaching the original NFMCD

Figure S.2
GLSC Is the Neutral Integrator for Execution

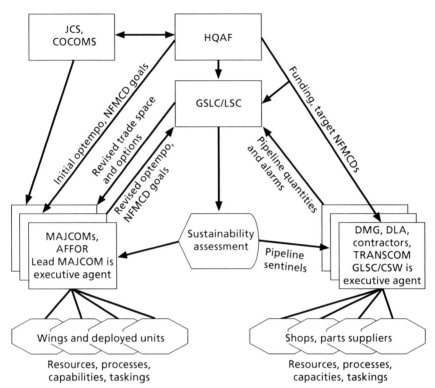

NOTE: The "GLSC/LSC" in the figure represents the Air Force's current transitional situation, in which implementation of the GLSC is still under way and, therefore, separate MAJCOM logistics support centers (LSCs) supplement the GLSC.
RAND MG667-S.2

goals for each force. When the disruption is larger, the GLSC would undertake a replanning effort in concert with the supply and demand side agencies that would rebalance the supply-side and the demand-side activities—for the near term. That revised plan might lead to changing the financial or other resources available on the supply side or changing the optempo or NFMCD goals on the demand side, depending on the circumstances and the overall military requirements.

Conclusions and Next Steps

As appealing as this conceptual DLR planning and control system design may appear, we anticipate that considerable refinement and extension will be needed for this design to be applied successfully. Certainly, any application of the new system design will encounter constraints and procedures that, while effective or efficient in the current planning and control system, may interfere with this system's conceptual foundations. Over time, those processes will need to be revised to enhance the planning and control system's ability to ensure the intended NFMCD levels efficiently.

Finally, the process used to define a COP for DLRs could be applied to other materiel sustainment activities and product lines. (See pp. 99–101.)

Acknowledgments

The authors of this monograph have had the good fortune to build upon the concepts and perspectives of many of our RAND colleagues. Foremost among those was the late C. Robert Roll, who introduced us to the concept of using neutral integrators to allocate scarce resources in circumstances where there is no market to assure responsiveness, efficiency, and overall effectiveness to competing demands. Likewise, we have benefited greatly from Frank Camm's decision-rights framework, as it clarified the ways that different agencies within the material sustainment system might be able to use the COP we envision. Finally, we appreciate the thought-provoking reviews by Edward Keating and John Folkeson, whose comments caused us to sharpen and streamline this monograph extensively from its earlier drafts. Taken together, we have had the good fortune to "stand on the shoulders of giants."

That said, we alone are responsible for errors and omissions in this monograph.

Abbreviations

AAM	Aircraft Availability Model
ACS	agile combat support
AFCAIG	Air Force Cost Analysis Improvement Group
AFFOR	Air Force forces
AFMC	Air Force Materiel Command
ALC	air logistics center
AWM	awaiting maintenance
AWP	awaiting parts
CAM	Centralized Asset Management
CAMS/REMIS	Core Automated Maintenance System/ Reliability, Equipment and Maintenance Information System
CIRF	centralized intermediate repair facility
COCOM	combatant commander
CONUS	continental United States
COP	common operating picture
CSW	combat sustainment wing
DLA	Defense Logistics Agency
DLR	depot-level reparable component

DMAG	Depot Maintenance Activity Group
DoD	U.S. Department of Defense
eLog21	Expeditionary Logistics for the 21st Century
EXPRESS	Execution and Prioritization Repair Support System
FAD	force activity designator
FYDP	Future Years Defense Program
GIS	geographic information system
GLSC	Global Logistics Support Center
HQAF	Headquarters, U.S. Air Force
HUD	head-up display
INW	in work
JCS	Joint Chiefs of Staff
LSC	logistics support center
MAJCOM	major command
MDS	mission design series
MICAP	mission impaired capability, awaiting parts
MOE	measure of effectiveness
MOEI	measure of effectiveness indicator
MOP	measure of performance
MSS	materiel sustainment system
MW	maintenance wing
NAF	numbered air force
NFMCD	not fully mission capable, DLRs
NMCM	not mission capable, maintenance
NMCS	not mission capable, supply
NRTS	not repairable this station

NSN	national stock number
O&S	operations and sustainment
optempo	operational tempo
OIMDR	Organizational and Intermediate Maintenance Demand Rate
POM	Program Objective Memorandum
RSP	Readiness Spares Package
TDD	time definite delivery
TRANSCOM	Transportation Command
UND	urgency of need designator

CHAPTER ONE

Introduction

The U.S. Air Force (USAF) operates a large, multiagency, multifunction materiel sustainment system (MSS) to support its forces' operations worldwide.[1] That MSS includes not only official, organic agencies, such as financial management, engineering, depot and base repair shops, and planners, but also other elements of a non-USAF supply chain that encompasses other DoD and contractor suppliers of maintenance, modification, transportation, warehousing, and supply services.

The USAF MSS is continually caught between two countervailing pressures: demands for increased efficiency and lower costs on one side versus demands for increasingly effective sustainment of combat operations and peacetime training on the other. Those pressures have intensified over the past two decades as the demands on operating forces have become less predictable, resulting in increased intra-USAF competition for available funds and making it increasingly difficult to forecast which resources and services will be needed to meet near- and long-term sustainment demands. In addition, force modernization plans intended to arrest the rise in equipment (especially aircraft fleet) ages have begun to further intensify the competition for limited USAF funds.

---

[1] By *sustainment* we mean all the activities associated with operating an Air Force system throughout its entire life cycle, including modifying it to increase its safety, efficiency, or effectiveness in its current missions and to extend its ability to conduct other missions. Under this definition, we include activities that expand a system's capabilities, not just activities that maintain its original capabilities. However, we explicitly exclude those activities that acquire those systems.

The coordination of such a large, multiagency, multifunction enterprise is difficult, even in a relatively stable environment. Ideally, every agency would act in synchronization to deliver just the right amount of materiel and services to meet the USAF operating forces' demands. With limited funds, that would require balancing and trading off all investments, expenditures, and activities in each of the agencies against the operating forces' planned activities and operational goals, then allocating just enough resources to the various sustainment agencies to meet those goals.

In an extreme steady-state scenario, one might imagine an enormous, centralized linear or nonlinear mathematical program that would decide what activities would be performed, when, and by what agency. The program's allocation decisions would seek to minimize total operating costs while achieving some specified operational activity levels and performance goals and recognizing detailed sustainment resources and process constraints. As this monograph will detail, we believe this level of centralized decisionmaking is beyond the capacity of current technology.

More important, the Air Force MSS is never in steady state, making detailed central coordination all the more difficult. The operating forces' ever-changing operational activity levels and goals compound the difficulty by adding an element of uncertainty about the future requirements and capacity to deliver sustainment. Thus, funds and other resources originally acquired and allocated for some future period may not match the actual sustainment demands that are driven by the forces' changing taskings and operational goals. Perhaps more troublesome, the time required for planning and allocating funds and other resources is so long that the planning assumptions may not reflect operational requirements or sustainment resource capacities when the funds become available.

Thus, the USAF needs a flexible materiel sustainment system that can adjust rapidly to changing operational activities and goals while maintaining the highest possible efficiency and effectiveness. As we will argue in succeeding chapters, a centralized mathematical optimization program is not the solution to this problem.

However, a fully decentralized decisionmaking system is not the answer either. Such a system may place too much emphasis on locally observed efficiencies and local goals, to the detriment of the overall system performance and military operational goals. What is needed is a way to ensure that the decentralized decisions are coordinated and adjusted in accordance with the changing operational goals and priorities while remaining within constrained resources to the extent possible.

In this monograph, we suggest that a more flexible, efficient, and responsive materiel sustainment system needs a common operating picture (COP) to help plan and coordinate the activities of all the related MSS agencies. Instead of providing detailed taskings for the individual agencies to ensure their close coordination and responsiveness to changing operational demands, a COP would help develop achievable, explicit performance goals, sentinels,[2] and diagnostic measures that each agency could use to judge its performance and detect systematic changes that might jeopardize achieving the overall MSS goals. Thus, the individual agencies would plan and adjust their activities to meet those goals, track and report their attainment by monitoring the sentinels, and use the diagnostic measures to rectify and work around performance shortfalls.

Scope of This Monograph

It would require a heroic undertaking to address in detail how common operating pictures might improve the coordination of all the various agencies and functions of the USAF MSS. Such a discussion would require more background about the many different MSS products and services that can fit within a single monograph and, moreover, would obscure the central notion of a COP. Instead, we have elected to rely

---

[2]   Sentinels monitor leading indicators that can be tracked to ensure that agencies are achieving their performance goals. They are called *sentinels* because, under the proposed COP, they would signal impending failure to meet the intended performance levels. When sentinels detect that something has gone wrong, the diagnostics would help identify the underlying cause.

heavily on a single example—the support of reparable components for aircraft—to demonstrate how a COP can be used to coordinate sustainment activities across multiple agencies in a dynamic, even turbulent, environment.

Organization of This Monograph

Chapter Two provides some historical background on common operating pictures, how one might exploit them to improve an organization's efficiency and effectiveness, and some limits to their exploitation. Chapter Three narrows the focus to the reparable component sustainment activity and outlines a framework for developing a common operating picture for a materiel sustainment system. Chapter Four describes an application of that framework to the specific area of reparable components. Chapter Five indicates some of the next steps that would be required if the Air Force were to implement this COP.

CHAPTER TWO

Framework: Common Operating Pictures, Effectiveness-Based Measures, Decision Rights, *Schwerpunkt*, and Nonmarket Environments

In this chapter, we review previous research on common operating pictures, identifying four related concepts that help frame how a COP may improve an organization's efficiency and effectiveness: effectiveness-based measures, decision rights, *schwerpunkt*, and neutral integrators. Common operating pictures have emerged as a central coordinating concept across lateral and vertical agencies in military operations. Effectiveness-based measures hold some promise for developing COPs for large, complex operations such as military sustainment. An emerging economic literature on decision rights provides some perspective about how COPs could be used to tighten the integration of the many sustainment agencies' activities. *Schwerpunkt*, which translates as "focal point," emphasizes achieving the commander's intent regardless of unplanned events that may render the original plan's details irrelevant, with even the lowest-level subordinates given wide latitude to take independent action to respond more quickly to those unplanned events instead of waiting for revised central taskings. Finally, we discuss an organizational framework to resolve conflicts between the suppliers of a service or product and the users when there is no market to set prices and ration scarce resources to the most valuable purposes. We briefly summarize each concept in turn.

Common Operating Pictures

Common operating pictures were credited with contributing to the rapid operational successes during Operation Iraqi Freedom in Iraq. For the first time, ground and air commanders at multiple echelons had access to coordinated maps that depicted the disposition of forces, the primary targets, critical terrain features, impending weather conditions, etc. Dispersed service and joint command centers in the Middle East, Europe, and the continental United States (CONUS) could all see the same up-to-date images that depicted both the current state of the conflict and the near-term operational plans in varying levels of detail. Commanders in the various command and control centers reported that the COP made it possible to understand how their own near-term plans would intersect with others and to respond rapidly to changing events on the ground, thereby enhancing the synergistic effects of the various forces' operations (Ackerman, 2002; Myers, 2002–2003). For the first time ever, most of the Army could answer three key questions with high confidence: Where are we? Where are our subordinates? And where is the enemy?

Just as important, the Air Force could share that same information. While the Air Force has always had relatively better information about their plans and fixed targets, this more dynamic, more accurate operational picture enabled it to improve its support to ground operations and to respond more rapidly to developing military opportunities.

The concept of a common operating picture for ground forces first emerged in 1997, as the Army sought to improve its own internal integration between their intelligence and operations branches (Glasser, 1997; King, 2000). While all the equipment and enabling systems were not fully deployed at the outset of Operation Enduring Freedom, the rapid successes in that operation led all services to acknowledge the importance of the COP capability and to accelerate or initiate efforts to develop their own common operating pictures.

The primary benefit associated with most operations-oriented common operating pictures implemented during and since Operation Enduring Freedom is to improve situational awareness, most often with databases integrated with a geographic information system (GIS).

By depicting the physical location of troops and other key battlefield entities, command centers can rapidly task their air and ground forces without providing detailed instructions, and the individual units can see how their operations interact with neighboring and opposing forces. More important, command centers can retask units as the situation changes, and the units can coordinate their interdependent actions more closely without burdening higher headquarters.

From these initial military operations successes, other government agencies have begun to develop their own GIS-oriented common operating pictures. In particular, the Department of Homeland Security has begun to develop and deploy its own COP capability to coordinate federal, state, and local agencies' responses to natural disasters and terrorist attacks. In situations in which the dominant concern is the physical movement of forces or materiels in light of distances, terrain, changing situational conditions, and physical capacities, a GIS-based common operating picture can help planners and operational managers synthesize a more comprehensive assessment of the entire situation than would be possible otherwise. Moreover, by having access to the same common operating picture, personnel executing those plans can respond more rapidly to a changing situation because they will not need to wait for direction from higher authority to take responsible actions.

Efforts to develop common operating pictures for other, nonoperational problems have not progressed as rapidly, though they have been credited with some successes. In particular, one logistician reported, "We didn't have an iron mountain, we had iron hills, because we could lower the number of days of supplies that we needed in theater because we had good in-transit visibility to the theater" (Cone, 2003). Other logisticians reported that intra-theater visibility still needs improvement (Dail, 2004).

In the sustainment COP designs to date, the vision has been to provide total asset visibility for supply items en route to units in theater,[1] so that theater and service logisticians can know where all supply items

---

[1] In Army parlance, "from factory to foxhole."

are, ensure that they are going to the right units, and redirect their distribution as needed.

While we believe that the services could develop a total asset visibility system to rival and extend commercial air express carriers' systems, we also believe that the proponents of such systems underestimate the decision complexity of managing such a system. It is one thing to look up the delivery status and location of a single item in transit and track its progress. It is quite another to project how all the items in transit will affect military readiness or ongoing operations; make decisions about how best to allocate the limited available assets for each of the many thousand items in transit; and allocate the likewise limited transportation and port capacities to best support those operations.

To help address that issue, we turn to the concept of effects-based metrics as a framework for understanding the broader implications of ongoing operations in a complex system, then we identify actions one can take to improve those operations selectively.

Effects-Based Metrics

The concept of effects-based metrics was derived from Air Force writings on effects-based operations. John Boyd (1986) argued that the psychological effects of military operations on an enemy's ability to continue operations were much more important than simple attrition of enemy forces. John Warden (1988) identified a hierarchy of target classes that, if attacked by air power, would have a greater effect on a war's outcome than simply attacking enemy forces. David Deptula (2001) explained how effects-based operations planning in Operation Desert Storm had more important effects on the Iraqi forces' ability to continue operations than the direct effects of attrition.

Of course, effects-based operations are more difficult to forecast, monitor, and measure than attrition (Davis, 2001). While many organizations develop effective measures of performance (MOPs) that indicate whether approved procedures are being followed, most organizations find it very difficult to develop operationally helpful measures of effectiveness (MOEs). This is because the effects may not be observable until long after the key operations are complete.

Take, for example, a business introducing a new consumer product. Research and development activities must be initiated, detailed designs and manufacturing processes developed, production initiated, and advertising distribution plans developed and executed—long before the first product is sold. It may be months or even years before the board of directors or the stockholders know whether that new product paid for those initial investments and made a profit.

Thus most effects-based organizations complement MOPs that ask "Am I doing this right?" and MOEs that ask "Have I done the right thing?" with measures of effectiveness indicators (MOEIs) that ask "How am I doing so far?"

To return to the business example, MOEIs may take the form of customer surveys, prototype mockups, and focus groups during the research and design stage, product design and testing during the early production stage, and comparisons of actual against forecast sales and costs once production commences. None of the MOEIs provides an indication of ultimate profitability of the product line, but they provide important feedback that can refine the product, sharpen the advertising and distribution campaign, reduce production costs, or even terminate the product line before more good money is sent after bad. MOEIs are important complements to MOEs in those circumstances under which an organization cannot directly or immediately observe the effects of individual decisions and actions. Thus, the MOEIs act as indicators that can signal whether the desired effects will or will not be achieved.

Another example of effects-based metrics is the implementation of the Balanced Score Card as a strategic management approach throughout the Air Force. In the materiel management arena, the Balanced Score Card reports not only on the overall MOE of aircraft availability, but also on MOEIs such as customer wait time that can signal a forthcoming deterioration in that MOE.

MOEIs can provide important information about dozens or even thousands of individual decisions and actions. Does that mean a COP that contains all the necessary MOEIs would make it possible and desirable to centralize all decisions in a single control center? After all, if there is a single operational picture: Why can't a single manager or control center make all the operational decisions? For the answer, we turn to the emerging economic literature on decision rights.

Decision Rights

An emerging area of study in the economics community is the allocation of decision rights to different entities within or across organizations (Bester, 2002; Garvin and Roberto, 2001; Jensen, 2005).[2] The central question that decision rights address is who (more specifically, which agency within an organization or, if an external contractual arrangement exists, which organization) should make each different class of decisions. Thus in our business example, the developers may decide the product characteristics, the board of directors whether to develop and produce the product, the marketing personnel the distribution and advertising plans, etc. Decision-rights analysts identify three relevant factors for the allocation of the right to make a specific decision:

1. information availability
2. decisionmaking capacity
3. appropriate incentives.

In the ideal circumstance, a single agency might have all three factors. Often, they do not.

As hinted above, it may be possible to create a COP that would include all the MOEIs about how well an organization is progressing toward its ultimate effectiveness goal for a particular class of decisions. If that is the case, why not centralize that decisionmaking? The answer lies in the three factors identified above.

---

[2]   See the appendix for a summary of the key elements of this research area.

Information Availability

Even if it has the MOEIs, a centralized agency may not have all the information needed to make the decision. In particular, economists call this *latent information*—information that is difficult to move, at least for time-sensitive decisions. Latent information may include some broad background knowledge of the particular process or activity being managed, fragile information about the current state of that activity and its resources or about other pressures being placed on the activity, or a comprehensive understanding of the activity's ultimate value to the end user. In those cases, it may be difficult to acquire the latent information in the central agency.

A critical piece of latent information is the diagnostic information necessary to determine what has gone wrong and what actions one can take to bring the system back into balance. While a central control center may be able to reallocate resources to mitigate some shortfall in a subordinate agency, it typically does not know enough about the underlying technical causes of that shortfall to fix it. Thus, a control center may be able to work around a problem, but the subordinate production center, because only it has adequate diagnostic information, is the only agency that can rectify the problem.

Decisionmaking Capacity

Another factor is whether a central control center has enough people and intra-center communications to decide all the detailed issues in a timely manner. If it has only a few people, they can easily experience information overload as they try to address multiple problems with multiple products at one time. If it has more people, there is the issue of how to coordinate their actions to achieve a synergistic effect on all the MOEs and MOEIs. That is, the central agency would need to develop a number of independent decision subagencies in the centralized location if the problem were too large, thereby recreating the poorly coordinated decentralized decisionmaking process and organization that the central agency was developed to rectify.

Appropriate Incentives

Of course, simply delegating decision rights to a subordinate agency with the appropriate information and decisionmaking capacity may run afoul of the appropriate incentives factor. If the subordinate agency does not have its incentives aligned perfectly with the overall organization goals, there is a risk of what the economists call *moral hazard*— that the subordinate agency will perform at less than the optimal level with regard to the overall MOEs or MOEIs in order to receive local benefits or avoid local discomfort.

Provided that the central agency is solving a fairly stable problem for which the situation changes slowly or not at all and that the problem can be decomposed into relatively independent parts, the standard bureaucratic solution of many embedded central subagencies can be very efficient. Each subagency can solve its part in parallel and ensure the intended outcome without affecting the other subagencies' solutions. However, if the situation changes rapidly and there is a need for rapid, effective integration of different agencies' activities, the traditional bureaucratic solution will not be able to keep up, in large meaure because it must process too much detailed information too rapidly to respond to the changing situation before that situation changes again.

How, then, can one allocate decision rights to ensure that continuing field operations can respond more rapidly to a fluid, even turbulent, situation yet still focus the field's energies toward achieving the intended effects? Enter the German Army notion of *schwerpunkt*.

Schwerpunkt

Boyd (1986) argued that the German Army derived *schwerpunkt* from the physical sciences as an organizational concept for enhancing the responsiveness and adaptability of large organizations. Its first widely noticed result was the operational concept of *blitzkrieg* during World War II. In physical science, *schwerpunkt* means the centroid, or center of mass, of an object. In its military context, it means the focal point or main objective or effect to be obtained in an operation or activity.

The idea of a focal point or a center of gravity has been understood for some time in military thinking (Clausewitz, 1832). Boyd extends that notion when discussing the *blitzkrieg* approach. In his view, the concept was broadened in those fast-moving operations to characterize the content of communications between a force commander and his subordinates. Implicitly, it was used to delegate broad authority to take independent action to those subordinates when unexpected opportunities or challenges arose. Rather than developing elaborate plans with specific actions and informing each subordinate about its role and taskings, the commander would communicate the overall objectives to his subordinates, along with individual responsibilities for each objective. Each subordinate would then review the feasibility of achieving his assigned objectives with his available resources, agree or request some adjustment in objectives or resources, then turn and allocate objectives to his own subordinates.

Thus, the communications at each organizational echelon down to the platoon emphasized what was to be achieved, not how it should be achieved. Individual commanders at all echelons were able to arrange—and rearrange—the assignment of subordinates' objectives as they saw fit. Thus, there was much more flexibility in the way that specific objectives would be achieved.

The main benefit of this flexibility was the ability of the individual commanders to react more rapidly to changing situations on the ground. Instead of requesting higher headquarters' permission to respond, they could take immediate action to exploit emerging opportunities or counter unexpected challenges. Thus, the *blitzkrieg* was exceptionally successful, even though the German Army often faced numerically superior forces.

Schwerpunkt enabled the German Army to respond to the uncertainties inherent in a fluid combat situation. While a traditional battle plan with excruciating detail would have sufficed if the enemy were not a cognitive adaptive entity, the inevitable uncertainties about the enemy's distribution or operational actions made it almost certain that any detailed battle plan would not survive the first few hours. By adopting *schwerpunkt* as a communication concept for controlling the forces, the German Army was able to respond more quickly than their opponents.

A key element in the concept of *schwerpunkt* is the notion of a shared frame of reference, i.e., shared concepts and values. Especially early in the war, every German Army officer shared a similar socioeconomic background and perspective. More important, they shared the identical military education from the German military academies and early training exercises. From a decision-rights perspective, this latent information enabled them to communicate in very succinct language both up and down the chain of command—indicating only what was unique about a particular situation or indicating the effect to be achieved.

Thus, this communications paradigm reduced the volume of detailed information flowing up and down the command channels, the distances that each communication had to travel, associated time delays before specific decisions could be made, and the amount of information that the command centers had to process. By delegating the "how" issues to lower echelons, the command centers could better focus on the more important issues of "what."

However, this arrangement has an inevitable internal conflict: inconsistencies between the desired objectives and the available resources. Infused with the "can do" spirit inherent in most effective military organizations, subordinates have a tendency to accept infeasible taskings from superiors. While extraordinary efforts can sometimes achieve amazing results with minimal resources, oftentimes they cannot.

On the other hand, subordinates may be reluctant to communicate their full capabilities, preferring instead to husband their available resources and retain them "just in case" something unexpected happens. This moral hazard may deprive other units of access to those resources, resources that might help them achieve their objectives.

Explained another way, there is a natural conflict between the supply side and demand side for any service or good. The demand side may demand more than the supply side can provide, or the supply side may not deliver all that it has available.

In many daily situations, one can rely on a market to resolve many of those conflicts and maintain a modicum of efficiency in the production and consumption of goods and services. In a market, competition among sellers (and also among buyers) assures that producers sell their products or service at prices near the minimum that sellers will accept; if the producers do not, other producers will enter the market and sell the product at a lower price, driving the higher-cost producer out of the market.

Of course, there is no market within a large, complex organization such as the Air Force. There may be several consumers of a particular product or service, but there is typically only one producer of a specific product. Adding to the complexity of the situation, the users within the Air Force (the squadrons, wings, and major commands [MAJCOMs]) and beyond the Air Force (the Air Force forces [AFFOR]) need a delicate balance of goods and services. This is especially true in our example problem of depot-level reparable (DLRs) sustainment (discussed in Chapter Four), in which a shortage in one DLR asset cannot be offset in any way by a surplus in another.

Achieving Efficient Resource Allocation in a Nonmarket Environment

The separation of the demand side from the supply side in a large organization enables each to concentrate on the technical aspects of their respective functions, leading to increased efficiency and effectiveness in these functions. However, that separation can lead to conflicts between those on the demand side (who care about the amount of product or service being delivered) and those on the supply side (who care about the feasibility and financial or other resource constraints on meeting all the demand side's desires). While much of that conflict can often be resolved by direct interaction between the demand side and supply side, as depicted in Figure 2.1, the conflict cannot be resolved through direct negotiation when the demand side's needs exceed the supply side's capacity to meet those needs. Under those circumstances, a neutral integrator is needed to step in and resolve the conflict, perhaps

Figure 2.1
Neutral Integrator Mediates Demand-Supply Conflict

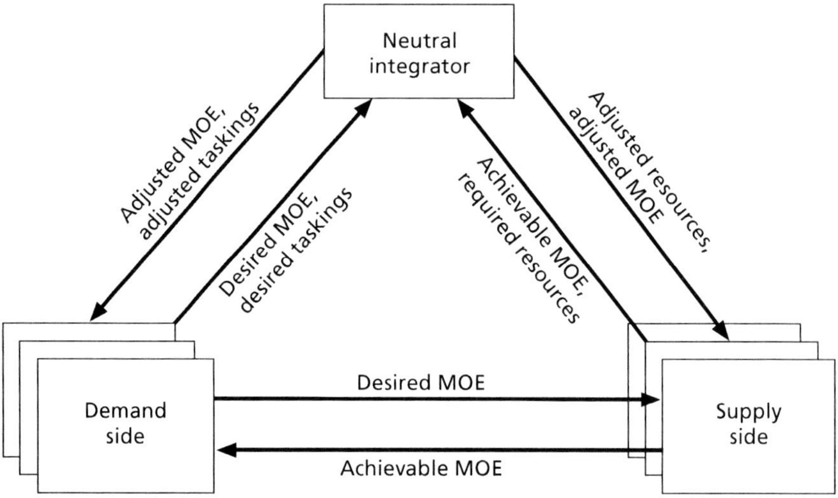

RAND *MG667-2.1*

by providing additional resources (often funding, but perhaps other resources as well) to the supply side, or by adjusting the taskings and expected performance levels for the demand side.

Note that Figure 2.1 also shows multiple demand-side agencies and multiple supply-side agencies. Thus the neutral integrator might balance a shortfall for one demand-side entity by rebalancing supply-side resource allocations in a way that limits the achievable effects for a different demand-side entity.

This iterative negotiation would continue until a feasible, operationally acceptable compromise was achieved. But the neutral integrator's role does not end with the completion of an agreed-on plan. Instead, the integrator also monitors the demand- and supply-side operations to ensure that the agreed MOEs are achieved and that the actual demands placed on the supply side are consistent with the agreed plan.

For that purpose, the neutral integrator requires the supply side to provide a series of MOEIs that could be used to monitor performance, along with a set of MOEI thresholds that, if violated, indicate that the system is not performing in a manner consistent with achieving the

MOEs. *Sentinels* can monitor these measures continuously to detect any unacceptable deviation from planned levels.

Should such a deviation occur, both the neutral integrator and the supply-side managers would spring into action. While the supply-side managers sought to identify the causes of the unacceptable performance and identify actions to rectify the situation, the neutral integrators would seek to identify potential workarounds, primarily adjusting resources and taskings, but also alerting the demand side of any potential shortfall.

Most often, the supply side will probably be able to rectify the problem before an MOEI reaches the threshold level or shortly thereafter. Should those efforts fail, the neutral integrator may have limited additional resources that can be brought to bear. In extreme circumstances, the neutral integrator may need to ask the demand side to adjust the MOE targets.

So far, this approach does not address the moral hazard issues. A demand-side entity can still demand more than it actually needs, and a supply-side entity can promise and deliver less than its actual capability.

Three solutions to the moral hazard problem present themselves: competition, experience, and computer models. If there are several equivalent suppliers or demand-side users available, the neutral integrator can use competition to ensure that the individual suppliers or users are delivering or demanding only those resources needed to meet the MOEs.[3] If the neutral integrator has access to personnel that have previous experience on the supply or demand side, he can draw on that expertise to evaluate proposed and actual performance levels, even if the users or suppliers differ in some ways. As a last resort, the neutral integrator may be able to use computer models of the suppliers'

---

[3]   Competition is probably easier to exploit on the supply side than the demand side. For example, there are often multiple suppliers of DLR repair and transportation activities. The neutral integrator could employ a periodic review of the sourcing decision to help motivate the current suppliers of those services to provide the requested services as efficiently as possible. There are fewer competition opportunities on the demand side, but several MAJCOMs share similar fleets, so it should be possible to compare performance and demands on those fleets.

processes to validate proposed MOEI thresholds and ensure that they deliver the required MOEs. In any case, the neutral integrator needs to have some independent means of verifying that the supply side is using its resources as efficiently as possible.[4]

---

[4]   Of course, there may be moral hazard on the demand side as well. We do not emphasize it in this monograph, but the neutral integrator also needs to verify that the various MOE objectives are really the levels needed, and that no MOE objective is overstated, implicitly limiting the resources available to meet other MOE objectives.

CHAPTER THREE

Method for Designing a Common Operating Picture

In this chapter, we outline one way to use the framework from Chapter Two to help design a common operating picture for a complex organization. Specifically, we envision an eight-step process that leverages the concepts of a neutral integrator and *schwerpunkt* to develop a set of COPs[1] and assign decision rights accordingly. In general terms, the process is as follows:

1. Identify the organization's broader objectives.
2. Relate those objectives to effects the supply sides must produce.
3. Identify measures of effectiveness.
4. Identify the processes and decisions that can affect each MOE.
5. Identify practical, comprehensive MOEIs and alarm thresholds for each MOE.
6. Assign the decision rights for the demand side, supply side, and neutral integrator to the lowest-echelon agency with the appropriate information and decisionmaking capacity.
7. Adjust the incentives for that agency to reward performance against the relevant MOEIs.

---

[1] We envision two different COPs for most materiel sustainment processes: one for planning, the other for execution. The planning COP will focus on the overall MOEs to be achieved, and identify boundaries that the MOEIs should not pass if those MOEs are to be achieved. The execution COP will monitor those MOEIs to assure that those boundaries are not violated, and it will act to bring the system back within bounds when needed.

8.   Periodically review these steps and adjust the COP and decision rights accordingly.

We discuss these steps more fully in the remainder of this chapter.

Identify the Broad Objectives

Almost all large organizations have a vision statement that identifies what they aim to accomplish. For businesses, it is a statement of a target profit; for nonprofit organizations, it may be the eradication of a disease or some social ill; for the military and other government organizations, it is usually one or more mission statements. Those objectives provide the foundation for developing a coherent management system that concentrates and coordinates an organization's efforts.

A large organization such as the Air Force is necessarily composed of many agencies with different technical specialties and local concerns. While a vision statement provides only general direction, with little concrete advice, it narrows the framework for the intra-organizational dialogs about what is to be achieved. While individual agencies may have their own concerns, the presence of a vision statement means that they must spell out how their concerns may affect the larger organization's ability to achieve its objectives, if they want to receive resources or other support from the larger organization.

Relate Objectives to Planned Effects

Of course, it takes more than a vision statement to motivate and channel the efforts of a large, diverse organization. To provide direction, such an organization also needs a plan or a roadmap to identify the specific effects it must accomplish to attain its objectives, as well as a time frame in which to accomplish the objectives. For businesses, that may take the form of annual revenue or cost-reduction targets. For a health-related nonprofit, it may require the development or promulgation of information regarding a disease's cause or cure. A plan or roadmap further concentrates the organization's efforts on those few activities judged most likely to achieve the overall objectives.

In this process, an organization sets individual agendas for its subordinate agencies' actions. Thus the marketing department may strive to increase sales, while the manufacturing department strives to reduce cost.

In most large, diverse organizations, there may be many potential plans that might achieve the overall objectives. A roadmap identifies to all agencies the single plan that they all must adopt. Thus, a sales program to increase the demand for one profitable product line must not conflict with a manufacturing cost-reduction program that would constrain or limit that product line's production to a level lower than planned by sales. Thus, both marketing and manufacturing need to focus their efforts on achieving coordinated effects, so that one department's effects do not limit those of the other department. The presence of a plan or roadmap is intended to assist in that coordination.

Identify Measures of Effects

To strike a balance across alternate effects, planners need concrete measures that gauge how different levels of individual effects may interact to achieve the overall objectives. Thus, businesses usually set specific sales and cost-reduction targets for individual product lines. Similarly, nonprofit health organizations may set specific targets for incidence rates and adverse outcome levels by target populations.

During this planning process, planners must use models ranging from the simplest "back of the envelope" estimates to computer models with highly differentiated and interdependent variables whose joint effects are not always intuitive. That is, they must make at least an educated guess about what will happen in the future, given their understandings of the supply and demand processes, the available resources, and different plans for allocating those resources.

To choose among the alternative plans, the planners and neutral integrators must judge which alternative plan's forecasted effects will best meet the overall objective. To do that, they must have some scales for comparing each alternative's effects against the others'.

Almost always, planners will need to balance multiple effects across all the plans to select the preferred plan. Assuming that dominant alternatives[2] rarely occur, the planners will need to perform tradeoffs across the alternatives to determine the preferred alternative.

When there are multiple conflicting effects that must be balanced in the face of limited resources, the optimum balance may not be computable in the traditional sense. That would only be possible if one had both a single utility function that linked all the various effects together and high confidence in the forecasts. If there were one single utility function that specified what the organization expected to achieve, it would be the only MOE needed. While businesses do have a single unifying utility function (profit), they do not have high confidence in the forecasts: Market uncertainties that they face in each of the components of that function make it necessary for them to develop contingency plans and hedging strategies that may not be optimum for how market conditions actually develop.

Thus, the neutral integrator would use forecast MOEs to evaluate alternative plans against alternative forecasts of future events, or scenarios. Implicitly, alternative plans assume different resource allocations and process performance levels for the productive processes on the supply side—plus anticipated demand levels from the demand side. Forecasts would need to use models (perhaps only crude ones) that relate resource-demand characteristics (frequency and size), resources, and process characteristics (timelines) to the future values of MOEs.

Once the neutral integrator has determined an appropriate balance of feasible MOEs, those balanced MOEs become a target level of effectiveness that the rest of the organization must strive to achieve.

---

[2]   A dominant alternative would outrank all other alternatives on all measures of effect. While such an alternative would vastly simplify the decision process, they occur only rarely in the real world.

Identify the Processes and Decisions That Affect Each MOE

Decisions are implicit in each and every demand- and supply-side process. The demand side chooses activity levels that generate demands. The supply side chooses which of those demands it will honor, how it will honor them, and how quickly it will honor them. The neutral integrator monitors their mutual progress and chooses when and how it may intervene to rectify conflicts or performance shortfalls.

Thus, the MOEs reflect not only the effects of physical processes associated with all the contributing agencies and activities, but also the effects of the agencies' daily decisions. As plans are executed, each agency chooses to use its productive resources in a particular way that may or may not contribute to the MOEs. Even though an agency has an appropriate level of resources and effective processes, its performance against its MOEs[3] also depends on its use of those resources and its refinement of its internal processes.

Identify MOEIs and Alarm Thresholds for Each MOE

Each effect being sought usually requires the joint action of multiple agencies within an organization. Each agency performs some particular activity that may consist of several different related processes.

Many things outside the supply agencies' control can intervene to interrupt even the best-laid plans. Demands can surge in unanticipated ways; resources can be diverted to other more important activities; processes that once served well can falter and become dysfunctional. Thus, MOEIs are needed to verify that actual processes are operating in accordance with the demand characteristics, resources, and process characteristics that were assumed while planning and setting the target MOEs. If something changes, the supply-side agencies need to respond or the target MOEs need to be revised accordingly.

---

[3]   We note in passing that an agency may have more than one MOEI. Its MOEIs should become, in effect, the agency's MOEs, and it should act to achieve the intended levels.

As discussed in Chapter Two, it may not be possible to observe the MOEs directly, except in very simple circumstances. First, it may take some time to achieve the desired effects, so one cannot expect the MOE to achieve its target value immediately. Also, it usually takes the joint action of many agencies in the organization to achieve the target MOE. If one or more agencies falter, it may be difficult to determine the cause and take corrective action. Finally, and of special importance to military organizations, it may not be possible to measure the MOE directly, because the demands of greatest concern do not occur except on rare occasions, such as major regional contingencies. In that case, the MOEs of interest could degrade below acceptable levels and not be apparent to the organization unless MOEIs are used to monitor the supply- and demand-side processes.

MOEIs should be closely related to the demand characteristics, resources, and process characteristics assumed during the planning process. However, they are not necessarily the same as those parameters. That is because the MOEIs need to be leading indicators of developing problems that might affect the MOEs. In particular, demand and process parameter changes can take some time to measure. Instead, the MOEIs should measure some value that can signal the change as early as possible.

Variation in the MOEIs over time is inevitable. Smaller variations may have little or no effect on the MOEs, so it should not be necessary to take corrective action every time a small deviation occurs. Rather, one is concerned with two basic situations: when an MOEI has passed some level that is "out of the ordinary," and when the MOEI's recent values reflect a trend that would jeopardize the target MOE if it continues. Thus, one can envision two kinds of MOEIs: *threshold values* and *trend measures*. Trend measures provide an important complement to the threshold values, in that they may signal a deteriorating situation before the threshold value is reached.

To ensure a system is achieving (or will ultimately achieve) its target MOE level, we envision using sentinels that automatically monitor the system's MOEIs. By comparing those MOEIs to threshold values needed to achieve the target MOE, the sentinel could signal operational suppliers, demanders, and neutral integrators of current or impending problems that would jeopardize their objectives.

MOEIs should be composed of measures easily obtained within the organization. Ideally, those measures would already be used to guide daily decisions. If the necessary measures do not yet exist, it will be necessary to construct them.

Allocate Decision Rights to the Lowest Possible Echelon Agency

At this point, we appeal to the notion of *schwerpunkt*. Under that concept, the decision rights for decisions in a rapidly changing environment are "pushed to the edge" of the organization. With the advent of modern information systems and a common operating picture, it is now possible—and desirable—to delegate a much wider range of organizational decision-rights options. While increased information flows might support a higher degree of centralized decisionmaking than in the past, they also support moving the detailed, real-time decisionmaking outward to the very edge of a large organization. Such a delegation of authority makes it possible for the organization to respond more rapidly to changing operational conditions (Alberts and Hayes, 2003).

But of course that presumes a shared frame of reference and incentives among all the decisionmakers. Thus, some decisions, such as choosing overall organizational goals, may require latent information, such as the operational value of different effects. It may be difficult to delegate those decisions to the very edge of the organization. However, it should be possible to identify acceptable levels (i.e., alarm levels that sentinels could monitor) for the MOEIs that are consistent with those organizational goals, then use those MOEIs to guide daily execution and monitor execution agencies' operations.

Thus, it is not possible to delegate all decision rights to the lowest possible decisionmaker. That would lead to anarchy, confusion, and ineffective operations. So where should one assign the rights for specific decisions? Students of the decision-rights literature would suggest that decision rights be assigned to the location where all the information is available, there is sufficient decisionmaking capacity, and the incentives are consistent with the MOEIs and MOEs being sought by the overall organization.

The Information Availability Matrix

We use an information availability matrix to allocate the decisions across different agencies and to evaluate situations for which a common operating picture could improve coordination across those agencies. Such a matrix describes the current information available to all the agencies that supply, demand, or oversee the allocation of resources for some specific set of decisions. For lack of a better term, we call those agencies "stakeholders," because they have a stake in the outcome of the process. That is, their performance will be judged by whether they achieve the overall organizational objectives. Because they are stakeholders, they also typically have some voice in determining the level of performance to be achieved, either demanders stating their needs, suppliers stating what they can achieve, or neutral integrators balancing needs against capabilities and directing the allocation of resources accordingly.

A notional information availability matrix is depicted in Table 3.1. In it, several discrete agencies are identified, with one column for each agency. Each row in the table shows whether a specific kind of information is available to each agency. At the far right of each row is a column identifying specific communications issues associated with each kind of information. For example, some information may be latent to a specific location (perhaps too detailed, too volatile, or unobservable elsewhere), or it may only be available as the knowledge internal to a person with some specific experience or education.

Each information availability matrix is devoted to only a single decision and the information that may be relevant to that decision. If an agency has access to the relevant information at the appropriate time, an "X" in the appropriate row and column indicates that fact.

In Table 3.1's greatly simplified notional example, we have chosen the decision to buy and repair spare parts for a single national stock number (NSN) during a fiscal year. The "Maintenance" and "Contracts" agencies are suppliers, the "Operations" agency is demanding

Table 3.1
Notional Information Availability Matrix
(Directly Observed Information)

Information	Agency				Communication Issues
	"Maint"	"Contr"	"Supply"	"Ops"	
Planned operations tempo (optempo)			√	√	
Total spare-part requirement					Unknown
Total spare-part budget					To be determined
Total maintenance budget					To be determined
Previous spare-part demands			√		Latent: data systems
Previous optempo			√	√	Latent: data systems
Maintenance capacity	√				Latent: shop manager
Maintenance budget					
Maintenance requirement					
Spare parts available worldwide			√		Latent: data systems
Spare-part condition worldwide			√		Latent: data systems
Spare-part replacement requirement					
Spare-part budget					
Spare-part price		√			Latent: fluctuates
Previous spare-part prices		√	√		
Previous support experience			√		Latent: item management specialist

sufficient spare parts to conduct operations, and the "Supply" agency acts as the neutral integrator to ensure that budget constraints are honored.

As one can readily observe, this initial information availability matrix is very sparse. That is because it represents the situation before any computations are made. If the system had no additional information, the decisions about how many of each NSN to purchase or repair would naturally fall on the supply agency, whose personnel have long experience with the historical demand patterns for the individual NSNs.

Of course, the Air Force has invested substantially in computational support that should significantly augment the supply agency's experience and intuition when forecasting the future need for spare parts. Without going into details in this illustration of the matrix, the Air Force has a suite of computational systems that compute the total spare-parts requirement and apportion that requirement to maintenance and supply, based on the available spares, their condition (serviceable or not), the projected optempo, and the historical relationship between optempo and demands for the individual NSN. Once that computation is complete, the supply agency has a great deal more (and more accurate) information available to help make the decision about how many of a particular NSN to purchase and how many to repair.

That computational suite also provides summary logistics factors that relate fleets' optempos to the financial requirements for DLR repairs and shelf stocks. Those factors enable the Air Force corporate structure to determine a total spare-part purchase budget and a spare-part repair budget, which is also communicated to the supply agency, as shown in Table 3.2.

Because total maintenance and purchase budgets rarely match their respective requirement calculations in size, the supply agency relies on its previous experience supporting various items to allocate the available repair and purchasing funds to different NSNs. As it refines its final decisions, the supply agency will also work with the maintenance and contracting agencies to ensure that the planned repairs and purchases of each NSN reflect the maintenance capacity constraints and the changing prices for new spares.

Table 3.2
Notional Information Availability Matrix
(After Requirements Computation)

Information	Agency				Communication Issues
	"Maint"	"Contr"	"Supply"	"Ops"	
Planned operations tempo (optempo)			√	√	
Total spare-part requirement			√		Estimated
Total spare-part budget			√		Determined by af corporate board
Total maintenance budget			√		Determined by af corporate board
Previous spare-part demands			√		Latent: data systems
Previous optempo			√	√	
Maintenance capacity	√				Latent: shop manager
Maintenance budget			√		
Maintenance requirement			√		
Spare parts available worldwide			√		Latent: data systems
Spare-part condition worldwide			√		Latent: data systems
Spare-part replacement requirement			√		
Spare-part budget		√	√		
Spare-part price		√			Latent: fluctuates
Previous spare-part prices		√	√		
Previous support experience			√		Latent: item management specialist

In this example, we show a "before" and "after" snapshot of the information available to make a decision for which there is already a viable decision process with some decision support computations in place. In a more general application, an information availability matrix would be developed at least twice, once in a descriptive analysis of the current ("as is") information availability, then again for each information distribution ("to-be") alternative that is under consideration. That is, it may be possible to redistribute the information differently so that the decision can be made more effectively. Alternatively, it may be possible to use modern computational techniques to create new, more highly integrated or more accurate information.

In the example, we were able to demonstrate how an existing requirement-processing system enabled a more accurate decisionmaking process. In the next chapter, we will suggest how adding a common operating picture may enable the organization to move some information to a more appropriate agency or even to multiple agencies whose efforts interact.

Of course, a common operating picture will not solve all information-distribution problems, because it cannot move latent information. However, when that latent information is embodied in a person, it may be possible to move the person to a different agency, thereby moving the information as well. We will examine both options in the more detailed analysis in the next chapter.

Adjust Incentives to Reward MOEI Performance

It may be impossible to fully rectify the mismatches between local decisionmakers' incentives and an organization's stated objectives, but it should be possible for the organization to create and administer rewards to individuals or other entities that contribute strongly to those objectives. That is, it may be difficult to remove entirely the influence of "significant others"[4] outside the organization, but it may be possible

---

[4]   In their review of attitude theory, Fischbein and Ajzen, 1975, report that the anticipated opinions of "significant others" is the key impediment to taking action based on personal

to shift the balance between those external influences and the organization's objectives.

The specific details of such incentives are beyond the scope of this monograph, but they could include direct rewards for specific performance levels or less direct rewards such as continued or increased long-term workloads for successful agencies. In the case of individual employees, it could include promotions or other long-term incentives. In the case of external contractors, it could include the development of long-term relationships that would enhance their confidence in future workloads. It would be incumbent on the larger organization to become one of the "significant others" that individuals use as a benchmark for judging their actions.

Periodically Review and Adjust the COP and the Decision Rights

Organizational change is inevitable. The demand side's needs will change; new supply-side technologies will emerge; the neutral integrator's strategic views will evolve. Just as important, the organization's understanding of its own processes will be refined as it continues to interact with the external world. New information will emerge, decisionmaking capacities will change, the supply-side processes and available resources will change, and incentives will shift. Thus, the MOEs, the MOEIs, the decision rights, and even the incentives will need to be adjusted over time to reflect these shifts.

---

beliefs and attitudes. Thus, an individual may believe that a given action would be "a good thing," but fail to act accordingly, because a spouse, relative, or friend might disapprove.

CHAPTER FOUR

An Example: Common Operating Pictures for the Materiel Sustainment System

This chapter has two purposes. First, we illustrate how one might apply Chapter Three's methodology to design a common operating picture and the organizational relationships it supports. Second, we develop a conceptual foundation for applying the methodology to the full range of U.S. Air Force materiel sustainment activities that affect aircraft—ranging from component maintenance to aircraft depot maintenance to aircraft modifications.

Thus, as this chapter discusses each Chapter Three step in turn, it proceeds like a funnel—first describing wide-ranging issues that affect all sustainment activities early in the chapter then successively narrowing the context, first to depot-level reparable components, then to the activities associated with planning the maintenance and resupply for those components, then finally to the daily actions needed to operate the DLR sustainment subsystem.

Identifying Broad Air Force Objectives

In Chapter Three, we suggested that the first step for developing a COP would be to identify the organization's broader objectives. Fortunately, the Air Force has recently documented those in an Air Force Roadmap (HQAF, 2006) that identifies agile combat support (ACS) as one of the six capabilities[1] that it is striving to develop in support of the nation's National Defense Strategy as further developed in the 2006 Quadrennial Review (DoD, 2006). As described in the Roadmap, ACS encompasses 26 different functional specialties, ranging far beyond traditional logistics concerns to include acquisition, personnel, health services, science and technology, and even chaplain services.

In the Roadmap's vision, those specialties all contribute to six ACS effects:

- a ready force
- a prepared battlespace
- a positioned force
- an employed force
- a sustained force
- a recovered force.

Of course, those effects reflect activities that create forces ready for operations, develop ACS command and control, establish operating locations in anticipation of deployments, deploy and redeploy the forces, protect the forces and generate operational missions, support the missions, and recover the force when the mission is complete. As the force (or a portion of the force) moves through the phases of force creation, preparation, deployment, employment, sustainment, and recovery,[2] different functions within the ACS system will play critical roles.

The sustainment of depot-level reparables is critical to all six phases. They support the maintenance of aircraft and other equipment that enables the training of aircrews during force creation. Their future support during contingencies requires posturing (locating) maintenance facilities that can ensure continued, responsive support to worldwide operating locations, many of which cannot be determined in advance. They must include sufficient serviceable spares to support

---

[1]   The other five Air Force capabilities are air and space superiority, information superiority, global attack, rapid global mobility, and precision engagement.

[2]   In practice, only a portion of the force will be in any given phase at a particular time. It takes time to create a force, prepare the battlespace, etc. For example, it is not possible to deploy the entire force simultaneously because of transportation resource constraints.

initial high-intensity operations after deployment. Also, their support system must provide the resilient, durable maintenance and resupply activities required to continue combat or other deployed operations over a prolonged period. Finally, they must recover quickly from peace-time or contingency deployments in order to be ready to respond to future contingencies and other deployed operations. While there are many other functions required for agile combat support, DLR sustainment is omnipresent.

Aircraft Materiel Sustainment Measures of Effectiveness

Next, we turn our attention to the second step in the analytic approach, which is to relate those objectives to effects the supply side must produce.

The first issue one must resolve in this step is to identify the supply side and the demand side. In some measure, the entire Air Force is the supply side, because they provide the Roadmap's six capabilities that support military operations during contingencies. That is, they seek to deliver those six capabilities to the combatant commanders (COCOMs) during contingencies and other military operations.

As a practical matter, the COCOMs cannot devote much attention to the intra-service sustainment systems, concerned as they are with the application of the joint forces to achieve broad military objectives. Each service provides force components that are intended to be sustained by intra-service resources and sustainment activities during contingencies and other operations. In the Air Force, those force components are organized into Component numbered air forces (Component NAFs) and, in special cases, Component MAJCOMs (USAF, 2006). While operational taskings may be determined by the COCOMs, the taskings are carried out by the Component NAFs/MAJCOMs, who, in turn, place demands on the Air Force's MSS during contingencies and peacetime deployments. Thus, there is a hierarchically nested array of suppliers, where the Component NAFs/MAJCOMs are suppliers of forces to the COCOMs but are the demanders of support resources and activities from the MSS.

As a further complication, the Air Force's major commands and their subordinate units are also the demand side during two of the operational phases, force creation and force recovery. That is, the Component NAFs/MAJCOMs may drive ACS demands during contingencies or other operations, but the MAJCOMs drive those demands as they strive to organize, train, and equip the forces for future operations. Indeed, during peacetime, the MAJCOMs and their subordinate units are the largest demander of MSS support. Even when some contingency operations occur in one theater or another, the forces not engaged will still continue to demand materiel sustainment, including DLR sustainment.

Thus, the MSS within the Air Force is always a supplier. The Air Force MSS encompasses several functions—maintenance, supply, transportation, engineering, and acquisition—that focus on the "equip" activity across all six operational phases. Not only does it acquire the combat and combat support equipment needed to deliver those capabilities, it also maintains, modifies, and replaces that equipment.

Thus, the MSS can be viewed as a supply chain, encompassing not only the original acquisition of major and minor equipment, but also the maintenance and supply activities at bases and depots, the sustaining engineering, the equipment modernization and modification processes at depots and contractors, the contractor support to existing equipment, and even the DoD agencies that supply transportation and distribution services. While that supply chain delivers literally thousands of different materiel and software products to individual units worldwide, its real end products are the equipment that support[3] the six capabilities identified in the Air Force Roadmap.

Thus, the MSS's effects are not limited to the numbers of each type of equipment acquired, but also to the modernization and condition of that equipment. As shown for aircraft in Figure 4.1, the portion of the Air Force MSS that focuses on aircraft fleets must balance

---

[3]   As suggested in the "organize, train, and equip" mission, equipment is only one element supporting each of the six capabilities.

five competing effects—operational suitability,[4] mission reliability, airworthiness, availability,[5,6] and cost[7]—by using modification or maintenance activities to counter the deleterious effects of changing threats, changing legal constraints, and progressive materiel deterioration. The dashed lines in the figure indicate that increases in the originating factors have a negative effect on one of the five measures, while the solid lines indicate that an increase would have a positive effect. For example, modifications such as enhancing on-board defensive systems often improve operational suitability, mission reliability, or airworthiness, but they also temporarily remove portions of the fleet from service (i.e., reduce availability) and they cost money to develop, test, produce, and install. For aircraft, the MSS must strive to achieve an appropriate balance among the five competing effects for each fleet over time.[8]

---

[4] Operational suitability reflects the characteristics of the aircraft design: whether an operational aircraft could carry out its assigned mission in the current operational environment. Mission reliability reflects the ability of a typical aircraft in a fleet to complete that mission, once it is begun. Airworthiness reflects whether the aircraft is safe to fly, regardless of its operational suitability or mission reliability. Availability reflects the fraction of an operationally suitable fleet that is available to begin a mission. Finally, operating cost measures the amount of money spent annually to ensure acceptable levels of operational suitability, mission reliability, airworthiness, and availability.

[5] Weapon system availability is a total measure of the number of aircraft available for operations averaged over a given period of time. Thus, it excludes time aircraft are not mission capable because of maintenance requirements (NMCM), not mission capable because of supply shortage (NMCS), or not mission capable because of both reasons (NMCB) and time aircraft spend in depot-level maintenance and modification.

[6] For the purposes of this monograph, we do not include sortie-generation activities as part of the MSS. While we acknowledge that a case could be made that generated sorties are the "real" effect as seen by the COCOM or MAJCOM, we envision the MSS as the system that equips the force, not one that uses the force for operations or training.

[7] Other systems will have different relevant operational metrics for sustainment. For example, availability of the Global Positioning System (GPS) is not as important as the ephemeris error (the error in the satellite's information about its own position).

[8] Of course, the MSS supports many other systems and operations, ranging from exotic space surveillance and tracking systems to more mundane aerospace ground support equipment and equipment needed to establish and operate airbases. The desired effects for those product lines may differ substantially from the desired effects for aircraft.

Figure 4.1
Air Force Materiel Sustainment System for Aircraft Fleets

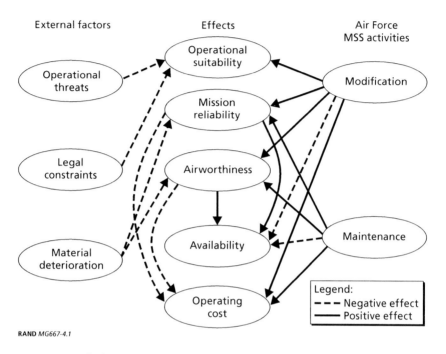

RAND MG667-4.1

One of the more troublesome aspects of the situation repre-
sented in Figure 4.1 is that there is a need for tradeoffs among the five
effects. Most conspicuously, the Air Force MSS must continually balance two countervailing forces: demands for increased efficiency and
lower costs on one the one hand versus demands for increasingly effective sustainment of combat operations and peacetime training on the
other. In addition, there are tradeoffs between increasing airworthiness (e.g., through more intensive inspections) and aircraft availability
(because the added inspections may take longer). Likewise, modifications remove aircraft from service, diminishing the number of aircraft
available for operations and training. Thus, there are tradeoffs among
these five effects for any given fleet.

Identifying Measures of Effects

Identifying effects is one thing; measuring them is quite another. Operational suitability is particularly vexing because it has so many dimensions. For example, an aircraft with no electronic countermeasures might be perfectly suitable to carry cargo in a benign environment. However, it would not be suitable if an adversary had capable radar-guided ground-to-air weapons deployed near a cargo operating location. On the other hand, that same aircraft might be operationally suitable for operations that involved an adversary with heat-seeking missiles, provided it had a flare dispenser.

Measuring Operational Suitability

Depot-level reparables have only a loose link to operational suitability. While modifications to DLRs can improve operational suitability modestly, the major improvements are accomplished through more comprehensive modifications that often replace many DLRs. Because we focus on DLRs in this example, we set aside the issues associated with measuring operational suitability. With suitable measures, the Air Force could add it to a future COP.

Measuring Mission Reliability

Things go wrong, not least with aircraft. Wear and tear, corrosion, fatigue cracking, electrical surges, and dozens of other phenomena cause complex equipment such as aircraft to fail. Some of the most troublesome aircraft failures are those that interrupt the successful completion of a mission, whether it be a fighter's combat mission, a transport aircraft's cargo delivery mission, or a reconnaissance aircraft's surveillance mission.

Failures can be categorized into three different types, depending on which type of aircraft components fail: DLRs, consumable spare parts (e.g., light bulbs, o-rings, or valves), or the aircraft infrastructure (e.g., surface skins, wiring, or hydraulic lines). Thus, DLR failures are only a subset of the factors that affect mission reliability.

To obtain an aggregate measure of mission reliability across all three sources, the Air Force collects two measures. Neither is a direct measure of mission reliability, but together they provide some bounds on that effect.

The first measure is the sortie abort rate. This measure tracks the percentage of times that an aircrew[9] cannot undertake an assigned mission because some mission-essential aircraft subsystem is not functioning properly. At that point, the aircrew is said to abort the mission.

The inverse of the abort rate is an upper bound on mission reliability. Whereas base maintenance personnel judged that the aircraft was functioning properly, the aircrew observed some problem (once power was applied or takeoff occurred) that the maintenance personnel could not detect. The equation expressing how the abort rate can be interpreted as an upper bound on mission reliability is

$$r_u \leq 1 - a \qquad (4.1)$$

where

r_u = upper bound on mission reliability, and
a = abort rate as a fraction of sorties attempted.

Of course, this is only an upper bound on mission reliability, because the aircrew could discover a subsequent failure during the sortie. The second measure addresses that subsequent failure by collecting the "Code 3 break rate." When an aircrew returns from a sortie, it reports either that the aircraft performed the mission flawlessly ("Code 1"), that some minor failure was observed that did not affect the mission ("Code 2," e.g., a torn seat cover), or that some failure was observed that must be rectified before any future missions are attempted ("Code 3"). The "Code 3 break rate" is the fraction of sorties in which a fleet's aircraft return in a "Code 3" condition. Taken in conjunction with

---

[9]   Airlift fleets track a related measure called the departure reliability rate. It records how often a fleet's aircraft can depart at the planned departure time, as a percentage of all airlift missions flown. It differs slightly from the abort rate in that it includes delays for prolonged maintenance that may preclude departing on schedule, whereas the abort rate only includes events that occur after maintenance has cleared the aircraft for operations.

the abort rate, the "Code 3" break rate can be used to estimate a lower bound on mission reliability, again as an inverse:

$$r_l \geq \left(1 - a\right)\left(1 - b\right) \qquad (4.2)$$

where

r_l = lower bound on mission reliability,
a = abort rate as a fraction of sorties attempted, and
b = "Code 3" break rate as a fraction of sorties flown.

Of course, this is only a lower bound on mission reliability, because it would include failures that were not discovered until after the mission was completed, or failures that did not affect the completion of the particular mission but might affect others.

Of course, the current measures include non-DLR failures, as discussed above. To assess the DLR sustainment system's effects on mission reliability, one needs to distinguish the DLR contribution to break rate and abort rate from these other contributing factors. With some small effort one could use the "action taken" codes from the Core Automated Maintenance System (CAMS) to compute the DLR-related abort rate and the DLR-related break rate as

$$a_{DLR} \approx \frac{\sum\limits_{WUCs} R_a}{S} \qquad (4.3)$$

$$b_{DLR} \approx \frac{\sum\limits_{WUC} R_b}{S} \qquad (4.4)$$

where R_a and R_b are removals after abort or breaks, respectively, S is the number of sorties conducted, $WUCs$ are the work unit codes included in the measure,[10] and a_{DLR} and b_{DLR} are the estimated abort and break rates for DLRs. If these values are substituted in place of the traditional measures in equations 4.1 and 4.2, they will yield upper and lower bounds on how the DLRs are affecting mission reliability.

A simpler, but less accurate, lower bound could be estimated from the organizational and intermediate maintenance demand rate (OIMDR) currently calculated by D200[11] to support the worldwide spare-part stock and repair requirement. Unfortunately, that measure includes removals unrelated to the abort rate or the break rate. Indeed, it includes removals of some items whose cumulative materiel deterioration patterns are fairly predictable or easily observed before failure (e.g., batteries, pumps, generators, etc.) and the pre-failure removals probably improve the fleets' effective mission reliability, because those near-failure items are removed before they can affect a mission. An equation for that simpler measure would be

$$r_l = \prod_i \frac{1}{(1 - d_i * l)}, \tag{4.5}$$

where r_l is the lower bound on mission reliability, d_i is each DLR's OIMDR, expressed in removals per flying hour, and l is the average sortie length, expressed in hours.

As with equations 4.1–4.4, this equation would be applied only to those components associated with mission reliability. While this simpler measure may include removals that don't occur as a direct result of

---

[10]   Specifically, this measure could be applied across the entire aircraft or to a subset of DLRs, depending on the scope of the decision being evaluated. In this particular case, we suggest that it should be applied to all the mission-related subsystems on the aircraft, not systems directly contributing to flying the aircraft. Those latter systems will be addressed in the airworthiness measure, discussed next.

[11]   D200, Air Force Materiel Command's (AFMC's) Requirements Management System, computes the annual requirement for investing in spare parts and repair actions to maintain an adequate level of spare DLRs in the field. To support that computation, AFMC tracks and records quarterly DLR removals by National Stock Number. When adjusted for flying hours, those removals can be used to reflect the probability that an aircraft will return from a sortie with at least one failed DLR.

failures during aircraft operation, it may be more sensitive to changes in the component reliabilities than the abort rate or the break rate. The larger sample size, which includes every removal for every component, may enable detecting DLR-related mission reliability changes before they affect the more subjectively measured abort rate or break rate.

Thus, one could then use equations 4.1–4.5 to estimate upper and lower bounds on the number of successful sorties in a forthcoming contingency or other operation. If so, one could use these measures to detect component-reliability changes and order by rank the value of redesign efforts to improve their reliabilities—based on the operationally relevant measure of overall mission reliability.

Measuring Airworthiness

Airworthiness is the technical term for an aircraft fleet's inherent flight safety, that is, setting aside human errors, such as aircrew or maintenance personnel errors, and setting aside other external factors, such as weather. As with mission reliability, there are many airworthiness concerns that lie outside the realm of DLRs and their support. The largest non-DLR concern is with unsafe aircraft structure and infrastructure, due to long-term materiel deterioration processes such as fatigue cracking, corrosion, and wire chaffing. Those concerns are managed through a comprehensive program of base and depot maintenance teams that periodically inspect the aircraft indications of such deterioration and remedy that deterioration accordingly.

However, aircraft flight control, cockpit indicators, navigation, propulsion, and aircrew escape systems include a large number of DLRs that may affect airworthiness. For example, the failure or miscalibration of an attitude indicator could lead to an incorrect indication to the aircrew and contribute to a subsequent accident. Of course, the most direct measure of airworthiness is the rate at which accidents occur. The Air Force counts accidents per million flying hours by mission design series (MDS) in three categories: A, B, and C. Category A mishaps are the most serious, in which one or more human lives are lost, $1 million or more damage occurs, or irreparable severe damage is done to the environment (AFMC, 2000). Categories B and C represent lower thresholds and exclude the higher categories.

Of course, Category A mishaps are generally lagging indicators because of their rarity. It may take some time before a sufficient number of mishaps of a particular type occurs for one to detect a pattern, certainly before one can say with some statistical certainty that a particular DLR or suite of DLRs may be contributing to the mishaps. While Category B and C mishaps occur more often, they are also best viewed as lagging indicators of airworthiness.

More important, the mishap rate measures include mishaps unrelated to airworthiness. By far, the major reported reasons for Category A mishaps are aircrew or ground crew errors. One cannot attribute such human errors to design or materiel condition issues inherent in the fleet, and these errors obscure emerging airworthiness problems. So it is exceptionally difficult to analyze the potential contributions of a DLR to a fleet's airworthiness.

Of course, abort rates and break rates will also be imperfect lagging indicators of underlying airworthiness issues in a fleet. Thus, one would prefer to rely on measures other than equations 4.1–4.4 to detect emerging airworthiness problems related to DLRs. Instead, equation 4.5 could be applied to the range of components judged essential to airworthiness, especially if one were to separate scheduled removals from the unscheduled removals that arise out of direct observations of materiel deterioration.

Measuring Availability

To be available, an aircraft must be both possessed by the operating commands (including test and training units) and capable of performing a mission within its designed capabilities. An aircraft can be possessed but not capable for two reasons—it needs a part or it is undergoing maintenance.[12] The Air Force declares an aircraft not mission capable (because of) supply (NMCS) if it needs a part that is not available on the base, and it declares an aircraft not mission capable (because of)

---

[12] Aircraft nominally associated with a unit are frequently sent to depot or contractor facilities for programmed depot maintenance or modification. Those "nonpossessed" aircraft affect the fleet's overall aircraft availability, but the aircraft are returned to the unit with the original DLRs (except where the depot may observe a failed component and replace it with a serviceable spare). Thus DLRs only rarely affect a unit's possessed aircraft.

maintenance reasons (NMCM) if either the aircraft or the part it needs are undergoing base-level maintenance. That is, an aircraft is NMCM from the time the maintenance technician begins to determine which part needs to be replaced until a replacement is provided from base stocks, the local component-repair facility repairs a replacement, or the facility declares the needed item "not repairable this station" (NRTS). Thus, DLR sustainment processes can contribute to either NMCS or NMCM in the standard Air Force measures.

As with the abort rate and the break rate measures, factors other than DLRs can contribute to the NMCM or NMCS status of an air-craft. Consumable parts such as limit switches, o-rings, etc., can cause an aircraft to be declared NMCS. Maintenance in response to a Code 3 break and scheduled maintenance inspections such as phases and isochronals[13] make an aircraft NMCM until the problem is resolved or the inspection is complete. So, NMCM and NMCS include the effects of factors and processes other than DLR sustainment.

To further confound the problem, some aircraft can be both NMCM and NMCS at the same time. Often, units "cannibalize" needed DLRs (and consumable parts, as well) from an aircraft under-going long-term maintenance such as a phase inspection. This reduces the total number of NMC aircraft, improving the unit's operational capability and flexibility. Such aircraft are usually coded NMCB (not mission capable, both) in the standard aircraft reporting system.

Yet another confounding problem is that a DLR's failure may affect only a subset of an aircraft's missions. Thus, an aircraft with a malfunctioning radar that would prohibit using the aircraft for a combat or long-range cargo mission can often still be used for less stressful sorties such as local familiarity flights, touch-and-go landings, etc. Such aircraft are not NMCS or NMCM but instead are coded par-tially mission capable (PMC).

Depending on the failed component's functional characteristics, a PMC aircraft may be usable for some combat or airlift missions. A

---

[13] Phase inspections and isochronal inspections differ mainly in the way they are scheduled. Phase inspections use the number of flying hours the aircraft has flown since its last inspec-tion; isochronal inspections use the number of elapsed days.

mission-essential systems list for each aircraft mission design series defines which aircraft subsystems must be fully functional for each kind of mission. For example, an aircraft may be capable of an air-to-air mission but not be capable of an air-to-ground mission under some circumstances.

To encompass the combination of all three measures, the Air Force has sometimes adopted the measure of not fully mission capable (NFMC). That is, an aircraft would be NFMC if it were NMCM, NMCS, PMCS, PMCM, NMCB, or PMCB.

To measure the effects of DLR sustainment processes on availability, one needs a measure similar to NMCS, NMCM, and PMC, but limited to DLRs. That is, one needs a single measure that encompasses the effects of both the base and nonbase component sustainment activities on aircraft availability. Thus, one would like to know the NMC and PMC caused by DLRs, without the confounding effects of other, non-DLR-related activities.

Because the DLR sustainment system is generally unaware of which missions are most critical at any given time, we argue that the DLR-related measure should reflect the all-encompassing scope of the NFMC measure. Thus, for measuring the effects of the DLR sustainment system on aircraft availability, we define the measure of not fully mission capable because of DLRs, or NFMCD.[14]

Of course, the definition is not the same as actually measuring the value. For that purpose, we turn to the worldwide stock control system that measures the number of components (including, but not limited to, DLRs) assigned by any unit, and the status of those components—serviceable, on-order, awaiting parts, or in local maintenance. For most components (those that are not contributing to NFMCD), the sum of the four status conditions will add up to the assets assigned to the unit. For those components contributing to NFMCD, the sum of those four numbers will exceed the number of assets assigned to the unit, and their potential effect on NFMCD can be measured as

---

[14] Over time, the Air Force may choose to adopt a multidimensional measure that captures the number of aircraft capable of each mission. We believe this initial measure will suffice for an initial COP.

$$NFMCD = \max_i \left(\frac{O_i + M_i + P_i + V_i - S_i}{q_i} \right) \qquad (4.6)$$

where

$NFMCD$ = number of aircraft not fully mission capable for DLRs at a unit,

$\max_i$ = maximum across all DLRs on the unit's aircraft,

O_i = number of component i on order,

M_i = number of component i due in from maintenance (DIFM),

P_i = number of component i awaiting parts,

V_i = number of component i serviceable,

S_i = number of component i assigned to the unit, and

q_i = number of component i installed on each aircraft.

Of course, this measure assumes that all missing components are consolidated (i.e., cannibalized) into the smallest number of aircraft possible. It represents the minimum number of NFMCD aircraft that a unit could achieve given the DLRs they have available and their current status.

Measuring Cost

All elements of a fleet's operations and sustainment (O&S) cost need to be included in any measurement of cost. This ensures that the apparent savings in one area are balanced against the increase in another area (e.g., moving base maintenance to a depot).

The Air Force Cost Analysis Improvement Group (AFCAIG) has defined a system of accounts that encompasses these costs. The Air Force Total Operating Cost system uses personnel data and base-level financial transactions to measure O&S costs and attribute them to specific aircraft fleets. The Cost-Oriented Resource Evaluator model uses optempo and resource utilization planning factors to estimate a squad-ron's future O&S cost.

The major AFCAIG DLR-related cost categories include the following:[15]

1.2	base maintenance personnel
2.2	base materiel consumption (for aircraft repair only)
2.3	DLR depot maintenance and replacement costs
3.0	centralized (off-base) intermediate maintenance
4.1.4	support equipment overhaul
6.1	support equipment replacement.

Through 2007, these costs are measured by calculating base level expenditures. As AFMC's Centralized Asset Management (CAM) initiative is implemented, costs other than base maintenance personnel will be measured in information systems related to that initiative.

Identifying Decisions That Affect Each MOE

For the remainder of this monograph, we set aside all but two of the five MOEs we have just discussed. While operational suitability, mission reliability, and airworthiness are important materiel sustainment MOEs, DLRs have their greatest effects on availability and cost. Future work will attend to other sustainment activities more directly associated with the other three MOEs.

Interestingly, most broad efforts to improve DLR-related availability (i.e., decrease NFMCD) across all USAF fleets will necessitate increased expenditures. That is, they will require additional investments or expenses that can be applied to make more DLRs available to the operational units. In the distant past, the primary focus was on acquiring sufficient DLR stocks to ensure relatively high confidence that a certain fraction of each fleet will be fully mission capable. More

---

[15] The numbers preceding each entry in the list are the AFCAIG codes corresponding to a specific accounting category.

recent innovations have included efforts to improve the process for maintaining and delivering serviceable spares to the operating forces. To get some idea of where those availability-improving investments or expenses might arise, it is helpful to consider Figure 4.2.

Figure 4.2 shows the various activities required to support DLR sustainment in some detail. Broadly, line replaceable units (LRUs—DLRs readily testable and removable from aircraft) are removed from aircraft, replaced by another item from base stock (if available), then the reparable LRU progresses through up to three levels of component repair, transportation, and distribution. To ensure that aircraft do not spend too much time waiting for a spare part, inventories are kept at base, centralized intermediate repair facility (CIRF), and depot levels, so that items can be issued soon after a demand is made by maintenance at each echelon.[16] If a component is not available in base stocks, it is immediately requisitioned from higher supply echelons.

Of course, all of those activities take some time. Figure 4.3 shows the standard times allowed for many of these processes for one typical component (an F-16 head-up display, or HUD) in AFMC's Recoverable Item Requirements Computation System (D200). The supply levels at each echelon are intended to provide sufficient spares to cover the demands that would occur while waiting for local repair or resupply from a higher echelon. Of course, this number is contingent on the reliability of the component and the speed of the component maintenance and resupply system. If there were one worldwide demand each day and a failed HUD visited every repair facility (base, CIRF, and depot), one would expect 94 components in the entire DLR sustainment system on a typical day.[17]

Those stock levels are computed with a two-to-three-year lead time, and they allow for a measure of safety stock meant to guard against modest changes in the reliability or optempo. When the levels

---

[16] As complicated as Figure 4.2 is, it does not include the supply process for shop replaceable units (SRUs) or consumable components that may be required to accomplish the repairs in the various maintenance activities.

[17] As a practical matter, a substantial fraction of HUDs are repaired at base level and returned to stock without visiting CIRFs or depot shops, so the number would be substantially fewer. In addition, there is not a demand each day for F-16C/D HUDs.

Figure 4.2
End-to-End DLR Process Flow Diagram

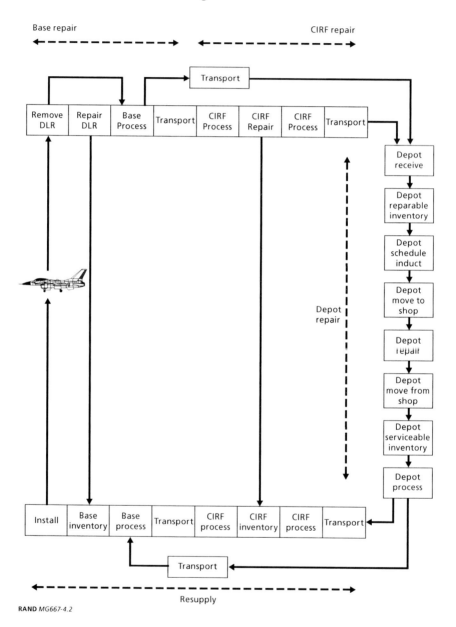

Figure 4.3
DLR Sustainment Process Nominal Flow Times (in days)

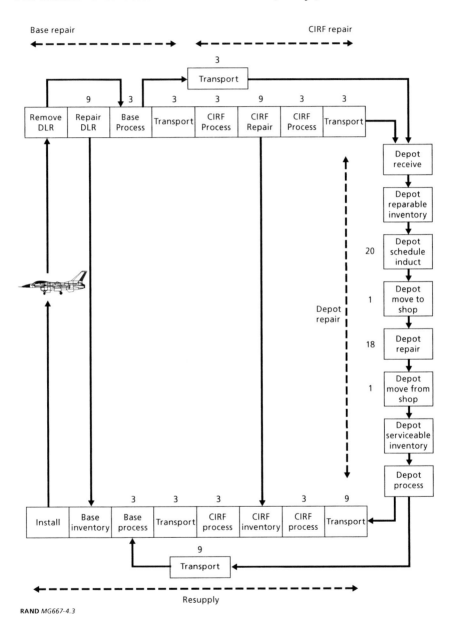

are too few to meet the demands, the available assets must be allocated to the individual bases to balance the potential sustainability shortfall in accordance with operational needs.

Five categories of decisions might cause the levels to be insufficient:

- funding for DLR spares and repair
- number of components acquired
- operational tempo (optempo)
- component reliability
- DLR sustainment system capacity and flow-time characteristics.

The DLR funding levels constrain not only the number of spares the DLR sustainment system can acquire, it also constrains how much maintenance can be performed to return failed DLRs to serviceable status. Very detailed, part-by-part acquisition system decisions affect the NFMCD level by allocating those funds to individual DLRs, striving to maintain balanced support across MDSs, across DLRs within each MDS, and between inventory and maintenance activities. In those balancing calculations, the acquisition system must assume specific optempos and component reliabilities in order to estimate future demands. If users choose different optempos to support operations and training, that balance will be imperfect at best and the achievable NMFCD level will change. If the users change other operating environment factors (e.g., changing the sortie length, the mission mix, the training syllabus, a fleet's configuration, or the forces' worldwide posture), many component reliabilities (per flying hour) will change and the balance across different DLRs will again suffer.

Finally, DLR sustainment system policymakers and decisionmakers are constantly striving to improve the system's efficiency by rebalancing transportation and maintenance capacity and flow times while simultaneously responding to changing demands within a limited constellation of resources.

There are different time horizons for those different investment decisions. Some decisions may require the entire Future Years Defense Program (FYDP—nominally six years) to implement (e.g., adding physical depot capacity), others may be implemented in a single year (e.g., increasing the financial resources available to repair DLRs), and still others may take a few hours or days to implement (e.g., adding a shift or changing repair and distribution priorities).

As a practical matter, DoD separates the longer-term financial-planning decisionmaking from the near-term decisions that implement or "execute" those plans. As part of their biannual planning, programming, budgeting, and execution process, every other year the services develop the FYDP, a comprehensive financial program that extends over the next six years. The FYDP is communicated in a document called the Program Objective Memorandum (POM). Detailed investment programs are proposed and approved in the POM covering the next two years, though limited adjustments can be made in the off year to reflect unexpected events. Each year, Congress reviews and authorizes the programs and appropriates funds for the following year's operations and support.[18] Once those funds are appropriated and the services receive the obligation authority, they can obligate those funds to execute their planned programs within the guidelines set by Congress. For this monograph, we consolidate the first three financial-planning activities into a single activity we call financial planning. Thus we distinguish between two decisionmaking activities and time horizons: financial planning and execution of those plans. By planning, we mean those decisions made about the FYDP during the POM development process. By execution, we mean those decisions made within the financial constraints created by the POM.

The key decisions made during financial planning include the following:

- force structure: what resources will be provided, including sustainment resources
- force posture: where those resources will be located
- fleetwide optempos (mainly for peacetime training)

---

[18] In this monograph, we are concerned only with the operations and support activities. Congress also appropriates funds each year for longer-term programs for which some materials or activities require a longer lead time to acquire.

- fleetwide contingency optempos (mainly planning factors for nominal major regional contingencies)
- cross-MDS NFMCD balance (implicit, not used in the latter stages of POM development)
- acceptable MDS-wide NFMCD goals (for MAJCOM/A4 internal use only, not used in POM development)
- depot and contractor repair and transportation budget
- base-level component shop personnel levels
- base-level shop equipment modernization and replacement
- component modifications or redesigns
- time definite delivery (TDD) standards based on Joint Chiefs of Staff (JCS) project code, force activity designator (FAD), and urgency of need designator (UND)
- cannibalization assumptions for peacetime (none) and readiness spare-part (full) calculations
- lateral resupply policy (none) for spare-part calculations.

Those decisions depend in turn on a number of long-term strategic decisions and operational characteristics of the DLR sustainment system:

- presence or absence of centralized intermediate repair facilities; CIRF locations
- base maintenance flow time, NRTS rate, and materiel requirement estimates by component (i.e., by National Stock Number)
- transportation time standards (both peace and war)
- depot and contractor flow time, hands-on labor standard, and materiel requirement and repair cost estimates by component (NSN)
- estimated component reliability (or the inverse, operations and intermediate demand rate, OIMDR), by NSN.

The key decisions made during execution include:

- actual force structure
- actual force posture
- actual optempo

- allocation of actual DLR spares and Requisition Objectives to units
- JCS project codes by unit and activity
- FAD by unit
- UND by unit and requisition
- actual cannibalization policy and actions
- actual lateral resupply policy and actions
- actual transportation capacity by channel
- base and CIRF overtime by base and shop
- depot overtime and multiple shift adjustments by shop
- contractor premium awards.

While these execution decisions do not explicitly consider NFMCD levels, they are key determinants of those levels. To the extent that they are not coordinated across the various decisionmakers, the DLR sustainment system may not achieve either the maximum NFMCD levels possible or the intended NFMCD balance across the multiple MDS and force users.

One way to view the two different decision classes is to consider that financial-planning decisions set broad, fiscally realistic, force-wide goals and constraints for the DLR sustainment system, whereas the execution decisions exploit that system's inherent near-real-time flexibility to reallocate funds and other resources within those budget constraints and also address the evolving sustainment challenges. This is an example of the principle of Boyd's interpretation of *schwerpunkt*, according to which broad policies and goals are set for the entire force well in advance of the actual execution, but decisionmakers closer to the actual events decide what specific actions will best achieve those policies and goals—within the available resources.

Today's Financial-Planning Decisions Imperfectly Address NFMCD Goals and Constraints

The individual fleets' NFMCD goals and the full range of DLR sustainment resources required to achieve them are not spelled out explicitly in the current financial process. That process uses pessimistic assumptions about the units' abilities to minimize NFMCD, it uses

NFMCD inconsistently throughout the process, and it fails to address how repair and transportation resource constraints will affect availability. In the very earliest stages of the FYDP development, the Air Force uses the Aircraft Availability Model (AAM) to estimate how many new and repaired spare parts will be needed to achieve a target peacetime level of NFMCD aircraft for each fleet, based on planned peacetime activities. In addition, it uses another model to estimate how many additional spare parts need to be kept in serviceable status in readiness spares packages (RSPs) for major contingencies.

The AAM calculations make a very conservative assumption about the ability of the DLR sustainment system to compensate for unplanned deviations in the demand processes—the calculations assume that maintenance personnel cannot move broken or missing parts among aircraft. In the real world, maintainers routinely cannibalize serviceable parts from aircraft already not mission capable to restore other aircraft to an operational status. As a result of cannibalization and other management actions, units routinely outperform the NFMCD levels used to compute the peacetime DLR spare-part and repair requirements. Models are available to reflect the effect of cannibalization policies on NFMCD; indeed, they are used routinely in the RSP calculations.

Today, complete calculations are performed only at the outset of the POM development process. Once that process begins, the subsequent programming decisions must be performed so quickly that there is no opportunity to revisit those calculations until well after the inevitable budget adjustments have occurred. To get some idea of how some optempo adjustments will affect the DLR sustainment costs, the financial planners use average "DLR costs per flying hour" factors derived from previous years' operations to estimate how changes in the training flying program will affect peacetime DLR costs, but there is little opportunity to adjust planned NFMCD levels during the rapid-fire adjustments required in the POM development process.

Finally, the underlying models for computing spares consider only the effects of spares and nominal flow times through a network that assumes unlimited repair and transportation capacity. Queuing models[19] can be used to estimate what additional resources would be needed to reduce the flow times, and the additional costs can be estimated from the cost of the additional resources. The resultant flow times can then be used as planning standards in existing models to estimate the spare-part and repair expenditures required to achieve the target availabilities.[20]

In summary, today's financial-planning decisions rely on imperfect models of the real-world support process. While some parameters, such as optempo and spare-part levels, are included, the overly pessimistic assumptions, inconsistent calculation techniques used, and the incomplete consideration of other resource constraints inhibit the Air Force's strategic and financial planners' ability to balance their DLR sustainment investments against other critical needs.

Many DLR Execution Decisions Also Address NFMCD Goals Imperfectly

Of course, many changes occur in the two to three years between the computation of the AAM and RSP levels and the actual execution of the final approved program. Not only do the available funds change, but so do the various fleets' priorities, the components' demand rates, and the available spare-part, repair, and transportation processes and resource levels.

Until fiscal year 2007, there has been no formal replanning process during execution to evaluate how to allocate the available funds across the repair enterprise in light of the changed priorities, evolving capabilities, and funds constraints. With the advent of Centralized Asset Management, the opportunity has arisen to rebalance the funds allocations across weapon systems and users. However, the only computational devices to support that rebalancing remain the AAM and the RSP calculations. Thus, any replanning will have the same imper-

---

[19] While those queuing models could be detailed, discrete models of each process, such detail is not required for the planning process. Loredo, Pyles, and Snyder, 2007, have demonstrated an effective, yet simple, approach to estimating arbitrarily complex shops' (such as a depot PDM shop's) throughput and flow time. That approach could be applied to DLR shops, as well.

[20] The computational details of one such approach are described in Hillestad et al., 2006.

fect resource allocations as the original planning and not reflect the actual processes and resource constraints.

Even the allocation of actual peacetime spares and requisition objectives to individual units does not reflect the changing plans. Instead, the authority to requisition spare parts is allocated to bases based on their historical use of those NSNs.

Further, the execution decisions are strongly affected by the external Joint Chiefs' assignments of JCS project codes and Force Activity Designators, but the execution decision rules implemented by the Air Force do not allow the DLR sustainment system to translate those decisions into NFMCD performance goals for the individual units involved.[21] An alternative system might recognize that it is well nigh impossible to restore the cannibalization aircraft at any unit to a fully mission capable (FMC) condition and set unit-specific NFMCD goals that would be used to declare more achievable NFMCD goals, even in high-priority units.

Finally, simulation research has shown that management adaptations such as cannibalization and lateral resupply can have a very strong benefit on NFMCD—far outweighing the effects of state-of-the-art demand prediction techniques (Adams, Abell, and Isaacson, 1993). Especially in execution, it is important to consider how the system actually operates when planning to allocate scarce resources to multiple operations and activities. Beyond cannibalization and lateral resupply, we would also include the adaptations that address the repair and

---

[21] The Air Force implementation of JCS project codes and FADs outlines a strict priority for distribution and transportation decisions by unit and urgency of need level. This can lead to sending serviceable DLRs to units that cannot make good use of the item (i.e., to fill a hole in an aircraft that needs many items before it becomes FMC), when sending the same item to another unit would make another aircraft FMC. An Air Force Corporate decision has mandated that those same priority rules be applied to repair production and induction in the Air Force Materiel Command's own organic component repair shops. While the Air Force has only limited influence over the JCS project codes and the FADs, it can affect how units state their UND, which should provide some control over the NFMCD that can be achieved by different units.

transportation constraints, such as overtime, multiple shifts, and priority decisions for component repair, distribution, and transportation.[22]

Identifying MOEIs and Setting Alarm Thresholds

We turn now to identifying MOEIs relevant to DLR sustainment. We will set aside the MOEIs that one might use to assess progress on operational suitability, mission reliability, and airworthiness, so that we can examine those that affect availability and cost in somewhat greater detail.

We envision MOEIs that form a DLR common operating picture that would be useful for both financial planning and execution. That is, the DLR COP would provide feedback about underlying process changes that affect how well the DLR sustainment system is meeting the intended NFMCD goals. Those process changes would inform the long-term financial-planning process for funding DLR spares, repair capacity and utilization, and transportation utilization by providing information about how changes in those resources would interact with operational plans (force structure, force posture, and optempo) to affect NFMCD. Those same process changes would be monitored closely by the execution system to detect changes that require reallocating repair and transportation capacities in near-real-time.

First we discuss the MOEIs that we would recommend, then we discuss setting appropriate thresholds.

MOEIs for DLR Sustainment Monitoring

Figure 4.3 (on p. 51) diagrams the DLR materiel flow network in some detail. As the figure shows, there are a number of different processes, each of which must operate consistently to ensure adequate aircraft availability. Some of those processes are mutually dependent (e.g.,

---

[22] The Air Force uses a system called EXPRESS (Execution and Prioritization Repair Support System) to prioritize the induction, repair, and distribution of many DLRs maintained at each of its own organic depot shops. However, there is no consideration of how allocating funds differently across shops might affect NFMCD rates. Rather, the shops are given a target "burn rate" to achieve during a given period.

depot DLR repairs depend on rearward transportation of the items to be repaired), and others may be complementary (e.g., base or CIRF DLR repairs may concentrate on items that rarely require depot activities, or vice versa).

Optempo, Flow Times, and Inventories

The most important observation that one can draw from Figure 4.3 is that changes in any one of a number of different parameters could have a devastating affect on aircraft availability. If the optempo, component removal rate, base repair time, CIRF repair time, depot repair time, inventory at any of those locations, or any of the transportation times among those locations were to change, the number of aircraft needing a particular part might increase to an unacceptable level.

Fortunately, any increase in the NFMCD caused by changes in any of those factors will take some time to occur. That means it should be possible to monitor the individual demand processes and the different process flow times to detect an unfortunate change before there is an unacceptable effect on NFMCD aircraft.

Nine Causes of NFMCD Changes

We identify nine factors that can affect a fleet's NFMCD rate: the fleet's optempos, components' reliabilities (i.e., demand rates), hands-on repair flow times, repair shops' parts inventories, repair shops' locations, hands-on transportation flow times, transportation capacities, component-repair shop capacities, and requisition priorities.

Some of these factors are under the Air Force DLR sustainment system's control; others are not. Specifically, optempos, components' reliabilities, and transportation capacity are not under the DLR sustainment system managers' control. Thus, it is critical for those managers to know about any change in those factors that might threaten the NFMCD goals, so that they can take remedial action to compensate for the change or at least minimize its effects on the intended NFMCD levels.

We first address the obtaining optempo forecasts. Then we turn to predicting future component demands. Finally, we discuss the ways in which the other seven factors interact to affect an aggregate flow time.

Monitoring Optempo

While future peacetime optempo may be relatively easy to obtain, contingency optempos may be more difficult. Both security concerns and the underlying turbulence of contingency operations make their optempo predictions less reliable. Nevertheless, the Air Force component headquarters have forecasts of future optempos that would be helpful in assuring adequate DLR sustainment to projected operations.

Most peacetime optempo deviations from the norm can be forecast well in advance. Many bases have local climate variations that affect peacetime flying, with shorter days characterized by reduced training operations. The Air Education and Training Command bases have seasonal fluctuations that depend on the training syllabus. Finally, most exercises are planned well in advance, and their optempos can be predicted with some confidence.

Monitoring Components' Reliabilities, or Demand Rates

Adams, Abell, and Isaacson (1993) evaluated several different procedures for monitoring how DLR demands have been changing and forecasting the implications of those changing patterns for varying forecast time horizons. One could apply their preferred procedures to substantially improve the accuracy of demand forecasts, even without addressing how different mission types affect demands.

Basically, they separated the worldwide components into three groups:

- parts with high worldwide demands (15 or more per quarter)
- parts with lower demands that cost less than $2,500
- parts with lower demands that cost more than $2,500.

Then, they evaluated the relative predictive accuracy of different models for each group. For the high-demand parts, they found that the most accurate forecasting model used both the current period's flying hours and the previous period's demands.[23] The most accurate model for the third group was the previous period's demands, while two models (the current flying hours *or* the previous period's demands) tied for most accurate model for the second group.[24]

Of course, it is the first group that should be of greatest concern to the Air Force DLR sustainment system. For any given percentage change in the demand rate, it is the parts in higher demand that will have the greatest effect on NFMCD. For example, a 50-percent demand increase for a component expected to experience 30 demands per period will create 15 unanticipated demands in the period; the same percentage increase for a component expected to experience 3 demands will have only 1 or 2 additional demands. While one might prefer to have no unanticipated demands, variations in the higher-demand components' reliabilities portend a greater risk to the NFMCD goal.

Thus, we recommend that the demand-monitoring process implement a partitioning of the components in a manner similar to Adams, Abell, and Isaacson, in which higher-demand components are tracked individually and lower-demand components are tracked as a group.[25]

Measuring Flow-Time Leading Indicators

The remaining seven factors—hands-on maintenance process time, repair shop location, maintenance capacity, spare-part inventories, hands-on transportation flow time, transportation capacity, and priority—collectively affect the system's flow times as depicted in Figure 4.3. Naturally, any increase in any of the flow times could cause a

---

[23] This may reflect some latency in the demand-recognition process. For example, some component failures may occur because of increased optempo in one period, but only a fraction of those failures may be detected by the maintenance system, causing some failures to appear as removals in a subsequent period.

[24] The authors note that it is not very satisfying to find that the most accurate demand-forecasting model was the previous period's demands, because that implies one does not know the causal process.

[25] The actual partitioning rules will vary with the fleet size and the period selected for monitoring. It may also be important to track some high-demand components' demand processes within subsets of fleets, perhaps treating MAJCOMs or deployed units separately. Resolving those technical issues will require further research and analysis.

subsequent increase in a fleet's NFMCD aircraft, at least for higher-demand parts. The number of aircraft needing parts might start at a "normal," stable level, then change only slowly until reaching a new steady state as the results of the longer flow time take effect. That is, it may take some time for enough demands to occur to attain the new pipeline steady-state level.

One might be tempted to monitor the individual flow times for each system segment directly. However, that turns out to be a lagging, not a leading, indicator of changing flow times because the usual measure for flow time involves recording the flow times for individual items as they leave the system. When a system changes, that after-the-fact measurement can only detect the change when the first item leaves the pipeline—well after the underlying change occurred. Ideally, one would want another, more responsive indicator that a change was underway.

The quantity of items in the pipeline is one such indicator. As soon as the flow time increases, the number of items in the pipeline begins to grow. Thus, one important indicator for all pipelines is the pipeline quantity, or the work in process (WIP).

But the beginning of pipeline growth is still not a leading indicator of potential capacity or hands-on process-time problems. At first, one could easily misinterpret the initial growth as just another random deviation caused by some demand or pipeline anomaly. Where possible, one would prefer additional direct measures of the factors that change those times: actual "hands-on" production time, capacity, and (for repair) location. We will return to address the issue of priority in the next subsection. First, we will discuss the MOEIs that one could use to detect changes in repair capacity and hands-on flow time.

Monitoring Hands-On Repair Time, Capacity, and Spare-Part Inventory Changes

When the repair workload exceeds the available repair capacity, some queuing results and some jobs wait for others to complete before they can begin (or in complex shops with multiple steps, continue to the next job step). If there were only one job in a shop, there would be no queuing and the amount of time required to complete that job would be the hands-on repair time. Such a job would be in an "in work" (INW) status for its entire time in the shop.

When there are multiple jobs under way, there is a chance that one or more jobs will encounter a queuing delay that will cause it to be placed in an "awaiting maintenance" (AWM) status for at least some portion of its time in the shop, thereby increasing the total time it spends there. As more jobs enter the shop, the probability of such collisions at each job step increases and the average job spends longer waiting for others to complete each step.[26]

Most DLR repairs require at least some new or refurbished materiel to restore the DLR to fully functional condition. Just as some spare DLRs are stored to cover the time while failed DLRs are undergoing maintenance, some materiels (both consumable and reparable subcomponents) are stored to be able to accomplish those repairs. Just as the DLR demands can exceed the available spare DLRs, the demands for those consumable and reparable subcomponents can also exceed available stocks, causing a DLR to begin "awaiting parts" (AWP). The AWP condition prohibits completing the repair until the missing component is received.

Some Air Force maintenance data systems, particularly the Core Automated Maintenance System/Reliability, Equipment and Maintenance Information System (CAMS/REMIS) and the standard base supply system in the base-level component-repair process, directly measure the time taken for each component-repair job. Likewise, they also directly measure the time in AWP status.

More important, they measure the number of items in AWM, INW, and AWP status. By aggregating items across NSNs to the maintenance shop level, one can detect whether some basic change has occurred. That is, a decrease in the INW with an increase in AWM might indicate some potential reduction in capacity, and an increase in AWP might indicate some parts shortages for that shop.

---

[26] In very complex shops such as engine repair or aircraft overhaul, queuing may only reduce the number of people working on an aircraft at one time. That can affect the labor "burn rate": the rate at which labor hours are actually applied to a specific task.

Similar measures are maintained at the depot component-repair shops for work in process and awaiting parts, though they are not as widely available. In addition, the depots' component-repair shops maintain a formal backlog of items that will not be inducted until both funds and sufficient capacity are available to commence work shortly after induction into the shop.

Thus, it should be possible to monitor the number of items in each of the three conditions (INW, AWM, and AWP) in a shop and detect or resource changes that may threaten the NFMCD goals.

Special Case: Repair Capacity

As indicated just above, the shops' pipeline contents may not react as rapidly to capacity changes as one might prefer. Thus, it should be possible to monitor two complementary MOEIs besides the pipeline contents: the number of direct personnel available, and the number of test stations in fully mission capable status. If either number declined, that would indicate that the shop might have difficulty sustaining an output level sufficient to meet the NFMCD goals.

The Air Force reports the status of some base-level maintenance equipment regularly in an equipment status report captured in the REMIS system. That existing status report could be used as a leading indicator for at least some key maintenance activities. Whenever the available (i.e., operational) equipment drops below some level, that would be a signal that the aircraft availability measure for the affected unit or units may be in jeopardy.

Currently, that system is limited to test equipment in avionics intermediate shops. It could be easily expanded to cover critical equipment in other shops, such as the engine test cells. Further, that system could be extended over time to cover equipment in depot shops. Such a comprehensive maintenance equipment status reporting system would constitute a large step toward developing leading indicators of maintenance capacity changes that threaten the sustainment system's ability to maintain acceptable aircraft availability levels.

Of course, maintenance equipment status is not the only capacity-limiting resource. Many processes have other resources that may constrain repair, particularly maintenance skills and materiels. Neither of these resource classes is currently reported, so it will probably be necessary to monitor the pipeline quantities, at least until reports on those resources can be developed.

Deliveries and Transportation Capacity as Leading Indicators[27]
Figures 4.2 and 4.3 vastly oversimplify the transportation system that delivers repaired DLRs to the operating units. In fact, even the CONUS direct deliveries from a depot to a unit require using three different modes of transportation.[28] Typically, the first is a ground transportation leg from a depot or contractor shop to an air hub, the second an air leg to a destination hub, and the third another ground leg to the unit. Any change in the resources allocated to each of those legs will cause a change in the capacity and response time for depot resupply.[29]

In practice, the ground leg flow times (from the supply action at the depot to the delivery of the item to the unit) may vary considerably, depending on the frequency of shipments from the source or to the destination. For example, a contractor shop whose total output can't fill at least half a truck daily probably would prefer to avoid the added costs of sending a mostly empty truck to the supply depot every day. Likewise, Transportation Command (TRANSCOM) would probably find other uses for a truck or an aircraft that is destined to carry only a handful of items to a small unit in a remote location.

Thus, the quantities in the transportation pipeline may vary widely between subsequent deliveries to a unit—even if the unit is get-

---

[27] With the regular use of domestic express transportation carriers for CONUS bases, one may have less concern about the issue of transportation capacity for CONUS bases. Thus these measures may be more useful for monitoring transportation support to overseas MAJCOMs and to theaters in a contingency. However, the USAF may also find it useful to monitor domestic ground and air delivery frequencies to detect any developing problems in that mode of transport.

[28] The exception is when the unit is collocated at the same base as the depot. For example, the 388 TFW at Hill Air Force Base, Utah, is collocated with the Ogden Air Logistics Center (ALC), which supplies many parts for its F-16 aircraft. Parts do not travel such a tortuous route between those two organizations.

[29] In the case of contractors performing DLR repairs, there may be four to six total legs: one to three from the contractor facility to the supply depot, and another three from the depot to the end user.

ting an appropriate amount of transportation capacity. One way out of this quandary is to track the average number of items delivered to a unit over some standard period (say a week or two). If the number of items delivered per period suddenly changed, it would be a signal that the transportation capacity dedicated to that unit had changed. To continue the example above, one would know at the end of the second week that deliveries had diminished substantially.

Given that the number of items or the flow time both would be subject to large variations, the preferred measure of the transportation system would be some more direct measure of the resources devoted to each leg or channel. For example, one might use the number of trucks or C-130 spaces allocated to a particular supply route per week.

While the total transportation capacity allocated to each channel would be sufficient for most items, some extraordinary items would need special attention. Classified components require special escort arrangements and asset-security arrangements. Hazardous cargo requires special political permissions when moved internationally and always requires special materiel handling processes. Engines are best moved by air ride trailers. Some very large items, such as C-5 engines, can only be carried by specific air or ground vehicles. While those items constitute only a small fraction of the number of items to be moved, they are a large fraction of the overall movement volume and weight.

Component-Repair Location

In general, it is often more efficient to move broken parts to an existing repair location than to move the repair location, particularly during contingencies when transportation resources are at a premium.[30] However, there are two circumstances when one might wish to consider moving repair capacity or at least redirect some repair to an existing facility: when either transportation capacity or a site's repair capacity

---

[30] The premier example of this situation is the movement of engine repair. While many engine repair fixtures are fairly portable, engine repair is performed most reliably in a clean, covered workspace with some overhead materiel handling equipment. Further, the activity requires a special engine test facility to verify that the repairs were adequately performed and that the repaired engine performs up to specification. Preparing such a facility on a bare base can require over 30 days.

is exceeded. In the first case, it may be necessary to move repair closer to the units to better utilize the available transportation capacity. In the second case, it may be necessary to redirect at least some reparable assets to a more remote site with available capacity.

In either case, one must track the available transportation and repair capacity and evaluate those resources' ability to meet the projected NFMCD targets, despite the unplanned imbalance between demands and the DLR sustainment system. In particular, if a capacity shortfall is projected to be fairly short (e.g., a temporary diversion of transportation capacity to another mission), it may not be necessary to change the repair locations, depending on the effect of that disruption on the NFMCD goals.

Priority Changes

As already mentioned above, DoD maintains a strict process for prioritizing the distribution of scarce materiel and for prioritizing movement of those materiels to all services.[31] That same priority rule is used in the Air Force's own depot shops to prioritize DLR repairs.[32]

Setting MOEI Alarm Thresholds

If the NFMCD goals are set with a fairly high confidence level (e.g., 90 or 95 percent), one can be reasonably sure that it will take more than a random variation from the expected levels of demands or flow times to jeopardize achieving the goal. That is, one can anticipate that some random variations will occur in the removal, repair, and transportation

---

[31] DoD's Time Definite Delivery system relies on units' FADs and the statement of urgency of need to allocate priority transportation and distribution actions among contending units. Changing a unit's FAD would have the effect of changing required transportation flow times for an item with a particular level of UND. It is even possible that increasing one unit's FAD would decrease the ranking of other, less well-positioned units' effective order and ship time, because items that would have gone to the other units would now be dispatched to the first unit.

[32] EXPRESS prioritizes induction and processing of reparable DLRs based on the TDD priority process. In addition, it prioritizes induction and repair of noncritical requisitions based on the probability of achieving unit specific NFMCD goals, based on the near-term optempo and a repair, transportation, and (where applicable) major regional contingency (MRC) surge duration.

processes that do not portend any lasting, systematic change in the underlying process. By setting the NFMCD goals with some margin for random error, one can avoid overreacting every time some minor, temporary dislocation occurs in the DLR sustainment system.

We begin setting alarm thresholds by identifying DLRs whose likely future demands have some "uncomfortable probability" that they will exceed the unit, MAJCOM, COCOM, or fleetwide NFMCD targets. In this test, we suggest using the Adams, Abell, and Isaacson forecasting approach to estimate the probability distribution of each item's demands that would occur in one complete circuit of the sustainment system. (Remember that this forecast also uses the near-term planned optempo, so it tests the changing demand rates and the optempo together.) By one complete circuit, we mean a weighted average of the various flow times for each item. This is equivalent to the steady-state pipeline quantity and the associated probability distribution.

Setting the actual "uncomfortable probability" threshold is a matter of some art, because there is a tradeoff between overreacting to false alarms and missing actual alarms. Fortunately, the large number of DLRs with very small demand rates makes it unlikely that most parts will ever have an important effect on NFMCD. Thus, it should be possible to identify a watch list of parts that have some potential to affect NFMCD, albeit fairly remote. While the low-demand parts need to be reviewed occasionally to ensure that they haven't migrated onto the watch list, the vast majority of demand-rate changes with some potential to affect the NFMCD level will come from items with fairly large demand rates.

Traditional approaches to monitoring items on such lists use statistical tests for deviations from nominal demand levels with probabilities in the range of 90 to 95 percent, representing the probability of mistaking a random swing as a significant event. Of course, the higher that threshold is set, the greater the chance of failing to detect an important demand rate or optempo change early.

Thus, we would suggest comparing the actual pipeline quantities against the current planning forecast quantities and their "uncomfortable probability" thresholds for only the high-demand components. That is, one would use the latest revised planning scenario including the planned optempo and component demand rates to estimate the Adams, Abell, and Isaacson probability distributions for the each DLR, summing up across all DLRs in the fleet for the each pipeline segment. Then the level of all DLRs actually in the pipeline at a given time would be compared to that level to detect any broad flow time change that affected all parts.

Adams, Abell, and Isaacson defined a process for improving predictions of how future optempo and recent demand levels interact to create near-term demands. Using that forecasting process, one could identify the DLRs whose demand probabilities pass the opposite threshold from the "uncomfortable threshold" used to test the total pipeline. For example, if one is using a 95-percent probability in the test of the overall pipeline flow time, one might feel comfortable identifying the "watch list" by selecting those items that have a 5 percent probability of surpassing the target NFMCD. Then one would watch those items' pipelines more closely, using traditional statistical tests to detect any change that placed them outside the probability threshold based on their nominal (planned) demand rate.

When Thresholds Are Exceeded

Should the demand, overall pipeline, or individual watch list items' measures exceed the thresholds, two parallel paths need to be pursued: one to discover and rectify the cause, the other to discover a workaround. Discovering the underlying cause may require little time or analysis, but rectifying it may take some time if additional or repaired equipment, additional training, or additional support materiel is required.

Workarounds would bridge the gap. In general, workarounds would reallocate currently available resources to minimize the overall effects of the DLR sustainment shortfall. In some cases, existing procedures such as cannibalization may absorb some of the effects on NFMCD. In other cases, the workaround may only require rebalancing support activities across all DLRs associated with a particular fleet or MDS. In a few cases, it may require reducing DLR sustainment to some fleets so that a critical capability provided by other fleets can be maintained at acceptable levels. Thus, these MOEIs serve as feedback to both the execution and the planning elements of the DLR sustainment system. While the execution decisionmakers work around some problem, the planners can arrange the longer-term process and resource adjustments needed to ensure appropriate support in the long run.

The Issue of Cost

With rare exceptions, the execution budgets for DLR sustainment are fixed. Thus, the execution decisionmakers operate under a rigid legal constraint that does not permit them to spend funds that were not authorized by Congress.[33] Execution decisionmakers may reallocate DLR sustainment funds across fleets or across DLR support activities in one fleet, but it will no longer be possible to spend funds this year and use future years' funds to recover the additional expenditures.

In extremis, it may be possible to obtain obligation authority from other programs outside the DLR sustainment area, but the execution decisionmakers will need concrete information to make such a case. Alarms based on the MOEIs suggested here would indicate that some planning assumption has been violated in a manner that affects the NFMCD goal for an individual fleet. Because those quantities are computed from planned demand rates, optempos, and pipeline flow times, one can use the measured quantities to estimate new values for those assumptions that could be used to develop a revised DLR sustainment plan, including the financial resources needed to maintain the original NFMCD goals. If those financial resources can be made available from another source, the revised plan could be implemented. If not, the plan would need to be revised further to estimate what NFMCD rate could be attained overall, then negotiations would need to be made with the MAJCOMs and components about which fleets will experience the lower (or higher) DLR sustainment level.

---

[33] This has not always been the case for the revolving funds used for DLR and some other depot support. The Air Force Materiel Command is implementing a new Centralized Asset Management system that is intended to enforce stringent topline annual budget discipline on the entire materiel sustainment system. Under that system, it will no longer be possible to perform sustainment activities without funds allocated from the current fiscal year's obligation authority.

Allocating Decision Rights

We turn now to the issue of who can and should make those decisions if a common operating picture containing the NFMCD goals and the MOEIs were available. Clearly, the decisions outlined above are being made now, but they are being made within a system that does not have a COP. It is useful to consider how those decisions are made today, and by whom.

Current DLR Planning Decision Rights and Available Information

Of course, it is the purpose of the planning activity to define the level of DLR sustainment to be achieved. Unfortunately, none of the planning decisions that affect the potential level of NMFCD even evaluate that performance parameter after the fact, let alone make any decision in light of the potential NFMCD tradeoffs.

As shown in Table 4.1, the closest the current system comes to considering NFMCD is in the initial D200 spare-part and repair requirement calculation of (no cannibalization) aircraft availability, based on nominal peacetime optempo, the worldwide fleets, historical demand rates, and average flow-time standards. That calculation is then used as an upper bound to guide financial programming decisions that develop the Program Objective Memorandum, which reflects the proposed Air Force budget for the coming five to six fiscal years,[34] during the Planning, Programming, Budgeting, and Execution process. That process has recently been streamlined to develop the POM in five to six fewer months (Gallagher, 2007).

Almost without exception, the actual funding allocated to DLR sustainment during the POM development process falls short of the level computed in the D200 process. In large measure, that is because the Air Force Programming process must balance many competing demands for funding and often must reduce the authorized training optempo below the nominal optempo levels assumed during the D200 calculations. To estimate the financial consequences of those reduc-

---

[34] For example, the fiscal year 2009 program that will be included in the President's Budget (PB) submission to Congress in February 2008 is based on the Air Force POM being developed in February–August 2007.

Table 4.1
Current DLR Planning Decision Rights

Decision	Decisionmaker	Information Used
Fleetwide optempo levels (peacetime)	HQAF	MAJCOM advice about required training, available O&M funding
Fleetwide contingency optempos	COCOM, adjudicated by SECDEF	Required operations
Cross-fleet NFMCD balance	HQAF	Available funds, $/flying hour factors, optempo and initial D200 calculation (i.e., no explicit NFMCD recalculation)
Acceptable fleetwide NFMCD goals	MAJCOM, within financial constraints	Historical NFMCD levels (i.e., no explicit NFMCD calculation)
Repair and transportation expenditures	HQAF	Available funds, $/flying hour factors
Pipeline flow-time standards (initial, peacetime, and war)	OC-ALC/CSW, OO-ALC/CSW, WR-ALC/CSW	Historical standards
Repair capacity	HQAF	AFMC advice
Spare-part investments	HQAF	AFMC advice
Planned component reliability	OC-ALC/CSW, OO-ALC/CSW, WR-ALC/CSW	Recent history, technical advice

NOTES: HQAF = Headquarters, Air Force; SECDEF = U.S. Secretary of Defense; OC-ALC/CSW = Oklahoma City Air Logistics Center Combat Sustainment Wing; WR-ALC/CSW = Warner Robins Air Logistics Center Combat Sustainment Wing.

tions, the Air Force uses DLR cost-per-flying-hour factors for each mission design series that permit a rapid reestimation of the required funding levels.

Unfortunately, the current method for estimating those factors is probably inadequate to ensure a given NFMCD level, because the demand process constantly evolves. For the last decade, units have not been permitted to requisition DLRs for which they do not have adequate funding, so they may have accepted a lower NFMCD level at the end of some years when actual demands exceeded the predicted levels. If so, the cost-per-flying-hour measure would underestimate the requisitions (and funding) needed to maintain the same NFMCD level.

(The cost-per-flying-hour factor is computed by dividing the cumulative exchange price[35] for all DLRs requisitioned by the cumulative flying for the same recent historical period.) Of course, the opposite effect would arise if the unit's funding were more than sufficient to maintain the initial NFMCD level, because units would strive to achieve the highest possible NFMCD. Not only is that measure poorly linked to the actual changing DLR demand rates, but it reflects an entirely different underlying variable—the DLR funding available to the units.

In 2008, the Air Force will introduce Centralized Asset Management (McCammant, 2006), which will remove the financial responsibilities for DLR funding from the MAJCOMs and units. In effect, DLR sustainment will return to a free issue system where cost data will be collected by AFMC and units' requisitions will no longer be restricted by the DLR funds they have available. Even so, the initiative is intended to adjust depot and contractor DLR inductions and expenditures to available funds during execution, regardless of changing demands (Kirby, 2007). If it is successful in that effort, the cost-per-flying-hour metric will still reflect the available funds, not the units' actual DLR demands per flying hour.

In either the current or the revised financial system, there is no way to ensure that a given level of NFMCD sustainment as program adjustments are incorporated in the POM. If the POM and the CAM are intended to ensure any particular level of NFMCD support, they will need to compute that value explicitly during POM development. A model could inform this process.[36]

---

[35] The exchange price is intended to be the net of the value of the broken asset and the original purchase price, accounting for inflation. As a practical matter, it includes both the average direct labor and materiel expenditures needed to repair the component, plus several maintenance and supply overhead and administrative charges. Prices for items with a replacement program underway also include a replacement charge.

[36] We note that non–Air Force agencies (the Office of the Secretary of Defense, the White House Office of Management and Budget, and Congress) make additional budget adjustments in the period between the POM completion and the allocation of funds to the Air Force during the annual budget. To the extent possible, agencies making those adjustments should also be made aware of the relevant operational consequences. Their understanding of those consequences will necessarily be limited by their experience.

Of course, those calculations depend on the characteristics of the demand and supply sides of the organizations. Currently, the demand rates and flow-time standards used in the D200 requirements calculation are based on a mixture of (mostly) standard flow times and (mostly) historically recent observed demand rates that are assembled by personnel in AFMC's three air logistics centers' (ALCs') combat sustainment wings (CSWs).

Ultimately, the NFMCD goals depend on how accurately the assumed D200 demand rates and pipeline time predictions, which are based on previous historical experience, match the actual demand rates and pipeline times that will occur in the near- or longer-term future. While the CSWs expend considerable efforts to monitor the demand rates and even adjust the forecasts of those rates based on information from technical experts, considerably less effort is devoted to adjusting the various pipeline time parameters to reflect changes in the DLR sustainment network. Thus, DLR sustainment planning has not been able to reflect the effects of different funding levels on NFMCD or any other availability measure. One can hardly imagine how it could be otherwise, given the very large volume of detailed information; its dispersion across multiple bases, MAJCOMs, and ALCs; the difficulty of collecting and revising the details for each DLR based on a combination of recent history and technical experts' advice; the huge decisionmaking capacity required to integrate the detailed information into measures suitable for adjusting the POM financial allocations; the limited decisionmaking capacity available; and the time pressures for rapid financial adjustments to the POM (on the order of hours, not days). In the past, computational power did not permit revisiting the lengthy D200 calculations more than two to four times per year, and only one of those calculations (the "March Scrub") is viewed as having authoritative, validated data.

Table 4.2 illustrates why this situation exists. All columns except the last represent the various agencies that have a stake[37] in the out-

---

[37] We view a stakeholder as an agency whose concerns may be affected by the outcome of a decision. Thus, a demander may find that a particular decision allocating DLR sustainment resources may or may not meet his needs. Alternatively, a supplier may find that same deci-

come. Those agencies include demanders, including the combat commanders, the Air Force component commanders, the Major Air Commands, and the units; the suppliers, including the commodity support wings, the MAJCOMs' logistics support centers, depot maintenance wings, the centralized intermediate repair facilities, the bases' repair and supply activities, and Transportation Command; and the potential neutral integrators, including Headquarters, Air Force, and the Air Force Materiel Command. The rows represent the information each may have that is relevant in the POM context. The last column identifies constraints on communicating that information.

Earlier in this monograph, we identified NFMCD as a key MOE that the Air Force would like to minimize. Indeed, Table 4.2 shows that MOE is not even calculated during today's POM process.[38] The main reason is that no one has all necessary information—the operators (COCOMs, AFFOR, and MAJCOMs) all have latent, hard to move, hard to share, knowledge about the value of higher optempos and additional aircraft in their particular context. However, they have no idea of the feasible operational tradeoffs in the face of financial and the DLR sustainment system's constraints. The CSWs, maintenance wings, contractors, CIRFs, bases, and TRANSCOM all have considerable information and data about the spares, repair capacity, and transportation capacity available, but it is widely dispersed. Those data are very detailed and voluminous, but they could be easily moved and shared with today's technologies. While the D200 process can relate the voluminous spare-part information to NFMCD, the only models of the repair and transportation capacity effects on process times are very cumbersome to use and demanding to keep current. Thus, it is beyond the DLR's current decisionmaking capacity to effectively estimate the repair capacity or transportation capacity needed to ensure the pipeline segment flow time needed for a given NFMCD level.

---

sion may or may not provide sufficient funding to meet those needs, support his work force, cover his costs, or overburden his available capacity.

[38] As discussed above, a related measure, (no cannibalization) aircraft availability, is calculated once, but it is not used for decisionmaking during the rapidly evolving POM development.

Table 4.2
DLR Sustainment Planning Information Availability

Information	COCOM	AFFOR	MAJCOM	GLSC	HQAF	AFMC	ALC/CSW	ALC/MW	Contractor	CIRF	Base	TRANSCOM	Communication Issues
Optempo (planned)													
Optempo operational value													Latent
NFMCD (planned)													
NFMCD operational value													Latent
NFMCD and funding tradespace													Capacity
Funding constraints													
Spares constraints													Detailed
Repair capacity constraints													Detailed[a]
Transportation capacity constraints													Detailed
Nominal demand, NRTS values													Detailed
Nominal pipeline times													Detailed

NOTE: MW = maintenance wing.

[a] May be latent for contractors.

Thus, there is little existing ability to assemble and process all the available information and turn it into an operationally relevant trade space to evaluate alternate funding levels or sustainment investments across fleets. The problem is just too complicated and time-demanding for the rapid decisions made during the POM development process, with thousands of different part designs per fleet and hundreds of organic and contractor DLR repair stations, many different transportation configurations and resource allocations, and with thousands of personnel.

The presence of two potential neutral integrators is a matter of some concern. At a minimum, they might hold two different views on the outcome, and any resolution of those differences will introduce delays in a very time-constrained process. That is, they each might envision a different balance in the DLR sustainment level across different fleets. Given the time it would take to rerun the D200 system with different goals, it would not be possible to recalculate the achievable NFMCD level within the POM decision cycle.

The fact that the final intended balance is not explicit also introduces the possibility that the intended balance will not be achieved. That is, the final obligation authority allocated to each of the depot shops through the AFMC chain of command may reflect the views and concerns of various decisionmakers within that chain instead of the balanced support envisioned in the POM. We now turn to discuss the execution of the financial plans developed in the POM and modified in the final budget approved by Congress.

Current DLR Execution Decision Rights

Even if the POM process settled on a target NFMCD rate for different fleets, the current DLR execution system would find it difficult to achieve those goals in practice. Table 4.3 summarizes the current allocation of decision rights for the DLR execution decisions outlined earlier in this chapter. As shown in the table, a fleet's optempo is rarely conditioned on the availability of DLRs. Rather, the optempo is driven by the need to train aircrews in peacetime and the operational requirements in contingencies. Cannibalization is managed locally by the unit, which tends to consolidate all mission parts into as few aircraft

Table 4.3
Current DLR Execution Decision Rights

Decision	Decisionmaker	Information
Actual optempo	Unit, COCOM	Operations and training requirements
Cannibalization policy and actions	Unit	MICAPs
Lateral resupply policy and actions	Unit, CSWs	MICAPs, DLR availability
Cross-contingency NFMCD levels by fleet	CSWs	Repair, distribution, and transportation priorities
Contingency versus operational NFMCD levels by fleet	CSWs, TRANSCOM	Repair, distribution, and transportation priorities
Cross-MAJCOM peacetime NFMCD rates by fleet	CSWs	Repair, distribution, and transportation priorities
Cross-unit NFMCD rates within MAJCOM and fleet	CSWs	Repair, distribution, and transportation priorities
Transportation capacity allocation	TRANSCOM	JCS guidance
Repair priorities	AFMC	FAD, UND, UMMIPS, quarterly workload negotiation, carcass and parts availability, EXPRESS prioritized list
Transportation priorities	TRANSCOM	FAD, UND, UMMIPS
Distribution priorities	DLA	FAD, UND, UMMIPS
Unit FADs	JCS	COCOM priorities
Requisition rules and UNDs	USAF, units	Supply Manual Chapter 24
Base, CIRF, and depot overtime and multiple shift adjustments	ALC/DMAG	Funding, workload availability
Contractor premium awards	ALC/SMAG	Funding, workload availability, contractor capacity
Replanning execution	ALC/DMAG	Funding, workload availability
Adjusted flow-time standards (near term)	N/A	N/A

NOTES: MICAP = mission impaired capability, awaiting parts; DLA = Defense Logistics Agency; UMMIPS = Uniform Materiel Movement and Issue Priority System; ALC/DMAG = Air Logistics Center Depot Maintenance Activity Group; ALC/SMAG = Air Logistics Center Supply Maintenance Activity Group.

as possible. Often, a unit may be able to reduce the number of aircraft missing serviceable DLRs, when they can locate that item at another base with similar aircraft. If so, that lateral resupply action may be performed through a base-supply-to-base-supply transfer, or the item may be directed to move by the responsible item manager in a commodity support wing.

The relative and absolute NFMCD levels are not used directly to decide what items should be repaired first, nor where they will be sent, nor how fast they will be sent. Rather, the units' achieved NFMCD levels are the result of the Time Definite Delivery system's rules to prioritize those activities' actions, described earlier. This means that some DLRs are sent to high-priority units that would not decrease those units NFMCD (because of aircraft missing other DLRs), while the same DLRs would decrease a lower priority unit's NFMCD.

The allocation of air and surface transportation capacity occurs entirely outside the Air Force. In contingencies, that allocation depends on the different combatant commanders' operational needs; in peacetime, on funding. In contingencies, at least some large fraction of the transportation system relies on DoD transportation assets. In peacetime, a greater percentage relies on traditional commercial carriers, especially for DLRs.

Even in the most generously resourced contingency, there is rarely sufficient transportation capacity to meet all users' needs simultaneously. As a consequence, the TDD priority scheme determines which items move faster, and to which units.[39] Likewise it determines which units' requisitions are filled first from available stocks and which items are repaired first in the depot.

In contrast to the transportation and distribution activities, whose primary limitations arise from their own resource constraints, the repair system is also constrained by the availability of broken carcasses to repair and the availability of spare parts (both bits and pieces and other DLRs) with which to accomplish the repair. Thus, it is quite

---

[39]   As a practical matter, most cargo destined for the same location is combined onto the same pallet. Thus, at least some lower-priority cargo enjoys faster movement to the high-priority units than its TDD priority might indicate.

possible for the repair system to repair lower-priority items when high-priority requisitions exist—just because the materiel and repair capacity exists.[40]

The NFMCD level is implicitly determined by the FAD and the UND during execution. While the FAD is assigned outside the Air Force, the rules for declaring a UND level are controlled by the Air Force and are documented in Chapter 24 of the Supply Handbook. The current rules permit units to declare the highest-level UND for any aircraft DLR for which there is no serviceable DLR in peacetime stock or in local base repair, even though receiving that asset will not increase the number of available aircraft at all (e.g., because of other cannibalized parts on the same aircraft).

The decisions to increase depot maintenance[41] or transportation capacity rely almost solely on the availability of additional funds and workloads, not on the level of NFMCD being achieved. That is, shops can work overtime and contractors can receive additional funding to work on a backlog, provided sufficient funds exist. In most cases, those additional funds come from outside the original DLR repair budgets, because diverting those funds from planned workloads would disrupt existing workload and labor allocations. Obviously, supplemental contingency funding is one source of additional funds, but other programs' funds may be diverted to DLR repair if a serious DLR sustainment problem arises.

Guiding DLR repair production based on available funding that is not linked to DLR sustainment goals by fleet can lead to serious imbalances in support within a fleet. Thus, some depot DLR repair shops can surge items for a few selected fleets while others surge for different fleets—leading to a situation in which no fleet's NFMCD rates are materielly affected because those repair surges were not synchronized.

---

[40] That practice cannot be considered wasteful, unless carried to extremes. The arrival of materiel for the higher-priority DLRs may allow their NFMCD levels to "catch up" with the levels that were achieved by repairing the lower-priority DLRs while the higher-priority DLRs were awaiting parts.

[41] As mentioned earlier, the decisions to temporarily increase base or CIRF DLR repair capacity are much less constrained by funding considerations. Rather, they depend on whether the shops are able to achieve a satisfactory NFMCD level.

There is no system in place to monitor, manage, or revise repair, distribution, or transportation flow times. Thus, while all the flow times in Figure 4.3 are used to compute the spares, there is no system to monitor whether those flow times are actually achieved. Instead, the only adaptation the system exhibits is to buy more stock or to induct more carcasses during a subsequent quarter.

In summary, it is remarkable that NFMCD goals have little direct influence on depot repairs, distribution, or transportation priorities or capacity adjustments. While that measure is commonly watched carefully at the base and MAJCOM, the TDD system that guides depot and other worldwide activities has only a crude indicator of NFMCD to guide its actions—the requisition UND, which often implies that meeting a high-priority UND demand will increase the unit's capability, even though no such outcome will occur when other missing DLRs prevent restoring an aircraft to a mission capable configuration.

Again, the lack of relevant information in the right hands is the main reason that the system has difficulty achieving the intended NFMCD level. Thus, Table 4.4 displays the limited information available to decisionmakers during execution. There are no MOEIs such as pipeline quantities or demand trends that one could compare with alarm thresholds to detect whether the system is operating as intended. Ultimately, there are only FAD and UND statements about individual requisitions that the MAJCOM logistics support center or commodity support wing can use to judge priorities and allocate scarce items. Worse, those measures are only available after the fact. The DLR sustainment system needs better indicators—indicators, such as those listed below the dashed line in Table 4.4, that can detect changes in system performance that will lead to unacceptable NFMCD levels.

Using the DLR MOEs and MOEIs to Guide and Monitor DLR Sustainment

As envisioned here, the DLR MOEIs and MOEs provide a common operating picture with which the DLR sustainment policy- and decisionmakers can improve DLR sustainment to the Air Force's aircraft fleets. We envision those MOEIs and MOEs as critical information that can support a closed-loop planning and execution system. As shown in

Figure 4.4, we envision an integrated DLR planning and execution control system that would allocate available DLR sustainment resources to various activities on the basis of the needed military capabilities. That process would take into account the operational capabilities of different fleets and the materiel sustainment processes and resources to allocate the materiel and productive (in this case, DLR sustainment) resources to those activities that make the greatest contributions to the Air Force fleets' capabilities. Specifically, we envision the planning process developing target MOEs (in this case, NFMCD levels) to be achieved for each fleet. During that process, there would be a need to identify the affordable cross-fleet MOE tradeoffs and decide what NFMCD level each fleet should receive.

Of course, that raises the question of who should decide. To answer that question, we turn to the twin notions of decision rights and the neutral integrator, first for planning, then for execution. But

Figure 4.4
A Closed-Loop Planning System

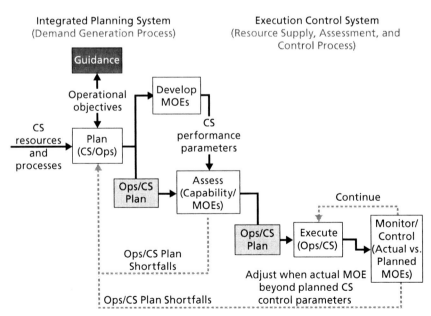

NOTES: Ops = operations; CS = combat support.
RAND MG667-4.4

Table 4.4
DLR Sustainment Execution Information Availability

Information	COCOM	AFFOR	MAJCOM	GLSC	HQAF	AFMC	ALC/CSW	ALC/MW	Contractor	CIRF	Base	Unit	TRANSCOM	DLA	Communication Issues
Optempo (planned)															
Optempo (planned, unit)															
NFMCD (planned, overall)															Capacity
NFMCD (planned, unit)															Capacity
Funding constraints															
Spares constraints															Detailed
Repair equipment available															Detailed,[a] Volatile
Repair personnel available															Detailed,[a] Volatile
Repair parts available															Detailed,[b] Volatile
Repair carcasses available															Detailed,[b] Volatile
Transportation capacity worldwide															Capacity
Transportation capacity theater															Detailed,[a] Volatile
FAD															Detailed
UND															Detailed,[a] Volatile
Demand alarm levels															Detailed
Pipeline alarm levels															Detailed

Table 4.4—Continued

Information	COCOM	AFFOR	MAJCOM	GLSC	HQAF	AFMC	ALC/CSW	ALC/MW	Contractor	CIRF	Base	Unit	TRANSCOM	DLA	Communication Issues
Demand levels															Detailed, Volatile
Pipeline quantities															Detailed, Volatile
Demand trends															Capacity
Pipeline trends															Capacity
Demand alarms															Capacity
Pipeline alarms															Capacity
Demand workarounds															Capacity
Pipeline workarounds															Capacity
Demand get well dates															Capacity
Pipeline get well dates															Capacity
Fleet alarms															Capacity
Fleet workarounds															Capacity
Fleet get well dates															Capacity

a May be latent for contractors.
b May be latent in CLS contracts.

first, we summarize an important technical computation development that makes it possible to integrate the effects of alternate funding levels on the NFMCD MOEs that can be achieved for each fleet during the tight decision cycles inherent in the POM development. We envision the results of that improved computation becoming a common operating picture used during the POM development to communicate the range of feasible, affordable DLR sustainment levels across fleets.

Further, we envision that the system parameters associated with the final financial allocation achieved during that improved POM development process will provide the information needed for more detailed measures that sentinels can monitor to detect when the MOEIs may be approaching unacceptable levels.

A Common Operating Picture for DLR Planning

As described above, the current planning system for DLRs is limited by the lack of a rapid process for determining how different optempos and investments in spares, repair, and transportation would affect the NFMCD level each MDS might be able to achieve. To address that problem, Hillestad et al. (2006) developed a capacity-sensitive computer model of the depot repair and distribution process that one can use to quickly examine a wide range of options. Their model addresses two key financial-planning and programming issues:

1. how adding funds to the spares or repair program would affect individual fleets' near- and long-term NMCS levels and contingency support levels
2. how moving funds from one fleet to another would affect those MOEs, while holding other fleets' support constant.

Given its rapid response time, the model could also be used to support decisionmaking beyond the POM deliberations and to replan DLR sustainment after the final budgets are adjusted by Congress.

Table 4.5 demonstrates how the new information content from the Hillestad et al. model could be made available to a wide range of DLR sustainment stakeholders to form a common operating picture viewed by all.

Table 4.5
How a DLR Common Operating Picture Could Improve DLR Planning

Information	COCOM	AFFOR	MAJCOM	HQAF	AFMC	GLSC/LSC	GLSC/CSW	ALC/MW	Contractor	CIRF	Base	TRANSCOM	Communication Issues
Optempo (planned)						New							
Optempo operational value													Latent
NFMCD (planned)		New	New	New		New	New	New	New	New			
NFMCD operational value						New	New	New	New	New	New		Latent
MFMCD and funding tradespace		New	New	New		New	New						Capacity
Funding constraints		New				New							
Spares constraints						New							Detailed
Repair capacity constraints						New		New					Detailed[a]
Transportation capacity constraints						New						New	Detailed
Nominal demand, NRTS values						New							Detailed
Nominal pipeline times						New							Detailed

[a] May be latent for contractors.

The table also incorporates an organizational change that is under way that would move more of the planning and execution responsibilities for DLRs to a new Global Logistics Support Center (GLSC). Under that concept, the MAJCOM logistics support centers and the ALC CSWs would be realigned organizationally under the administrative control of the GLSC, which would report in turn to AFMC. That administrative change is part of the much more comprehensive changes envisioned in eLog21 (Expeditionary Logistics for the 21st Century), a new Air Force initiative (Zettler, 2003; Bearing Point, 2005) that will provide new integrated information systems to guide and monitor the entire DLR sustainment enterprise from the bases to the depots and the contractors. The LSCs and CSWs will remain in their current locations, at least initially, and the GLSC will rely on virtual collaboration to ensure that the efforts of all parties remain synchronized.[42]

But the new information distribution across those agencies reflects more than just the results of using the Hillestad et al. model. To run that model, an agency must have access to all necessary data. Other agencies might be chosen, but we have selected the GLSC, because portions of that organization are collocated with the MAJCOMs and therefore may have access to latent information about operations that is not available elsewhere. In addition, those agencies include military personnel with previous field-level experience supporting operations and training activities who will certainly have latent information about how DLR shortages may affect those activities.[43]

While the GLSC might run the model, the headquarters AFMC, with its emerging CAM office, will serve as the neutral integrator, as shown in Figure 4.5.

The first step in this process would be to develop the trade space to provide information about a base case and how changes in funding

---

[42] See Wainfan and Davis, 2004, for a review of some of the challenges the dispersed agencies within the GLSC will face as they rely on videoconferencing, audioconferencing, and computer-mediated communications to synchronize and integrate their efforts.

[43] Of course, other agencies may be able to perform the same functions ably, provided they have personnel with similar experience and access to operations experts.

Figure 4.5
The AFMC Centralized Asset Management Office Is the Neutral Integrator for DLR Sustainment Financial Planning

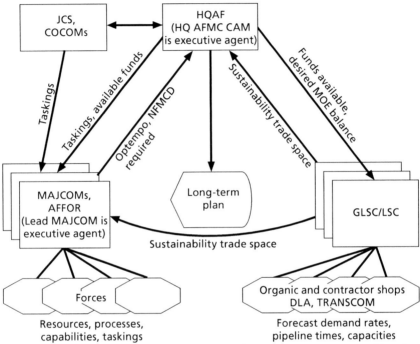

NOTE: The "GLSC/LSC" in the figure represents the Air Force's current transitional situation, in which implementation of the GLSC is still under way and, therefore, separate MAJCOM logistics support centers (LSCs) supplement the GLSC.
RAND MG667-4.5

for individual fleets would affect the NFMCD level for each MDS.[44] That would require assimilating information provided by the demanders (optempo) and the suppliers (demand rates, NRTS rates, and pipeline times) using the Hillestad et al. model. To ensure that the forecast is based on a solid empirical foundation, we recommend using the

---

[44] In some cases, such as the F-16C/D fleet, there are substantial block number differences within a fleet. The GLSC/LSC may find it necessary to develop a more detailed breakdown than MDS, so that operations experts can judge the operational value of different funding mixes. In some cases, the different uses of the same aircraft across different MAJCOMs may make it useful to develop the trade space, so that different commands may have different NFMCD levels for the same fleet.

experience of previous years' suppliers as the base case. That will ensure that recently achieved flow times reflect any queuing or other delays.

The next step would be to provide the boundaries of the initial NFMCD and funding trade space to the demanders (AFFOR and MAJCOMs). That is, the suppliers would provide the demanders with data that expresses how changing (increasing or decreasing) the funding would affect the NFMCD rate for each fleet. The demanders would then use those data to consider the consequences of funding changes for each fleet. Depending on the financial constraints, the neutral integrator (HQAF) might require the demanders to negotiate the NFMCD allocations with some fixed total budget in mind or allow the MAJCOMs to express specific needs that require funding over and above that initial level. In either case, the demanders could use the trade space of NFMCD versus funding for each fleet nearly instantly to examine the effects of different funding (by fleet and total) on the forecast NFMCD for each fleet.

If the original funding proved satisfactory to meet the demanders' requirements, the NFMCD goals would then be used to set alarm thresholds for monitoring the execution process, as described below. If not, the suppliers would need to revisit their analyses to verify that there is sufficient capacity available to meet the higher demand levels. That is, they would need to verify that the pipeline times originally used for the analysis can actually be achieved with the available repair, spares, and transportation resources—in the face of higher demand rates.

A variety of computational tools can be used to estimate the effects of changed workloads on DLR flow times for depot, CIRF, and base repair shops. These range from simple analytic queuing models for single-stage diagnose-and-repair processes to more general analytic queuing models of more complex repair processes such as PDMCAT (Loredo, Pyles, and Snyder, 2007) to still more complicated discrete simulations akin to the Logistics Composite Model.[45]

---

[45] The Program Depot Maintenance Capacity Assessment Tool (PDMCAT) uses simple mathematical calculations to estimate the dynamic production behavior of the very complex, multistage production processes inherent in aircraft Program Depot Maintenance, provid-

We anticipate that the physical capacity (facilities and equipment) will be sufficient in many cases but that additional labor and materiel may be required if workloads increase. Those factors, of course, are built into the cost-NFMCD trade space. That is, overtime or a second shift can enable the system to meet the pipeline times with little or no additional capital investments. If additional equipment or facilities were needed to meet the workload and maintain the planned pipeline times, the GLSC would identify the requirement for those capital resources and identify feasible workarounds (e.g., contractors, reduced optempo, or NFMCD adjustments) to bring the system back in balance. Those new workarounds (which should be few, if any) would be communicated to the affected demanders (through the neutral integrator), who would choose one alternative as an acceptable compromise until the required capital investment can be realized.

Computing the flow times through contractors' DLR repair shops would require information about those production processes, which may not be available to the Air Force in some cases. However, it should be possible to include a requirement in future contracts to provide that information or contractually binding throughput times for different workload levels.

Likewise, the Air Force may not have sufficient information to estimate how a changed distribution workload might affect the TRANSCOM flow times. This is especially true during major and minor contingencies, when transportation to or within a region may be disrupted or already burdened with other cargo. However, it should be possible to convert the demands into daily C-130 or C-17 equivalent loads by region and then ask TRANSCOM to estimate the delivery flow times.

Thus, the operational optempo, demand rates, and pipeline times become the foundation of a DLR sustainment plan for the execution year and beyond. Of course, for any number of reasons, these optem-

---

ing estimates on the number of aircraft in work and the annual production rate, based on relatively simple measures of total resources and production. The Air Force's own Logistics Composite Model uses a highly detailed simulation of the detailed sortie-generation tasks performance times and occurrence factors to estimate sortie production rates for a given array of maintenance personnel skills.

pos, demand rates, and planning flow times may not match performance levels achieved during execution. Therefore, we now turn to how one might monitor and adjust the execution of this DLR sustainment plan.

A Common Operating Picture for DLR Execution

During execution, the various DLR sustainment stakeholders need conformation that the system continues to operate as planned. When some part of that system deviates enough from the plan that it may ultimately threaten the attainment of the NFMC goals, one would like to detect the deviation early and determine the underlying cause as soon as possible. Thus it will be necessary to monitor the optempo, the demand rates, and the pipeline quantities, as outlined in the MOEIs described earlier in this chapter.

Also described there, it should be possible to set alarm thresholds around each of those parameters to detect when an unacceptable change has occurred that will ultimately jeopardize NFMCD, even though the current NFMCDs have not yet breached acceptable levels. Once those values have been computed, one can use the Adams, Abell, and Isaacson forecasting approach to compute demands, use those demands with the pipeline times to forecast likely quantities in each pipeline, and use the Adams, Abell, and Isaascson distribution to estimate confidence levels on those pipelines and set appropriate alarm thresholds.

At that point, the DLR sustainment system can use sentinels to monitor the MOEIs that may affect NFMCD. Initially, those sentinels may be people within the GLSC who periodically review the values of the MOEI for critical DLRs. Over time, we envision an automated system that could perform those monitoring tasks, alerting GLSC personnel when an MOEI was approaching or had surpassed its alarm threshold.

We envision a common operating picture would enhance that monitoring and reaction process. We expect the current information distribution across DLR stakeholders would change to resemble the one depicted in Table 4.6.

We envision the GLSC would compute, set, and promulgate the alarm thresholds throughout the DLR sustainment system. Over time one may find it desirable to set different alarm levels for different agencies in the material sustainment system. For example, individual base, CIRF, depot, and contractor repair shops might have explicit NSN-specific thresholds for each item they repair. TRANSCOM and the Defense Logistics Agency (DLA) might not have NSN-specific levels, but they would have sentinels that total quantity pipelines by location and channel. Those agencies would monitor their own pipeline quantities, in addition to the monitoring performed by the GLSC.

In particular, we suggest that the various repair and transportation (supplier) agencies monitor the trend values closely and detect many underlying problems well before they emerge as normal alarms. That is, we expect they would detect a potential trend early, extrapolate to determine when it might reach the alarm threshold or even affect some fleet's NFMCD goal, then take action to control that trend before it reaches either of those levels. This is possible because most shops only repair parts for a single group of similar commodities or a handful of fleets.

In contrast, the GLSC, serving as a communication conduit between the demand side and the supply side (see Figure 4.6.), would monitor only the MOEIs for all fleets. This is for two reasons. First, the individual shops may be able to detect and remedy many, or even most, problems (e.g., a repair equipment outage or a parts deficit) well before the NFMCD goal is threatened. Second, it is doubtful the GLSC will have the information-processing capacity to monitor and take independent action against every minor dislocation that may occur within the repair and transportation system. Rather, it must concentrate its information-processing capacity on problems beyond the capability of the individual repair shops or transportation providers.

Of course, the GLSC would monitor both the supply side and the demand side MOEIs, alerting the supply side if demands are expected to take a turn for the worse and alerting the demand side if supply is

Table 4.6
How a DLR Common Operating Picture Could Enhance DLR Execution

Information	COCOM	AFFOR	MAJCOM	HQAF	AFMC	GLSC/LSC	GLSC/CSW	ALC/MW	Contractor	CIRF	Base	Unit	TRANSCOM	DLA	Communication Issues
Optempo (planned)		N	N			N									
Optempo (planned, unit)		N	N			N	N					N			
NFMCD (planned, overall)				N	N	N	N	N	N	N	N	N			Capacity
NFMCD (planned, unit)						N								N	Capacity
Funding constraints						N									
Spares constraints						N									Detailed
Repair equipment available						N									Detailed,[a] Volatile
Repair personnel available						N									Detailed,[a] Volatile
Repair parts available						N									Detailed,[b] Volatile
Repair carcasses available						N									Detailed,[b] Volatile
Transportation capacity worldwide						N									Capacity
Transportation capacity theater						N									Detailed,[a] Volatile
FAD						N	N	N	N	N	N				Detailed
UND						N	N	N	N	N	N		N		Detailed,[a] Volatile
Demand alarm levels															Detailed
Pipeline alarm levels															Detailed

Table 4.6—Continued

Information	COCOM	AFFOR	MAJCOM	HQAF	AFMC	GLSC/LSC	GLSC/CSW	ALC/MW	Contractor	CIRF	Base	Unit	TRANSCOM	DLA	Communication Issues
Demand levels						N									Detailed, Volatile
Pipeline quantities															Detailed, Volatile
Demand trends						N	N	N	N	N	N				Capacity
Pipeline trends						N	N	N	N	N	N		N		Capacity
Demand alarms						N	N	N	N	N	N				Capacity
Pipeline alarms						N	N	N	N	N	N		N		Capacity
Demand workarounds						N	N								Capacity
Pipeline workarounds						N		N	N	N	N		N		Capacity
Demand get well dates						N	N								Capacity
Pipeline get well dates						N	N	N	N	N	N		N		Capacity
Fleet alarms		N	N	N	N	N	N								Capacity
Fleet workarounds		N	N	N	N	N	N								Capacity
Fleet get well dates		N	N	N	N	N	N								Capacity

NOTE: N = New.
[a] May be latent for contractors.
[b] May be latent in CLS contracts.

Figure 4.6
GLSC Is the Neutral Integrator for Execution

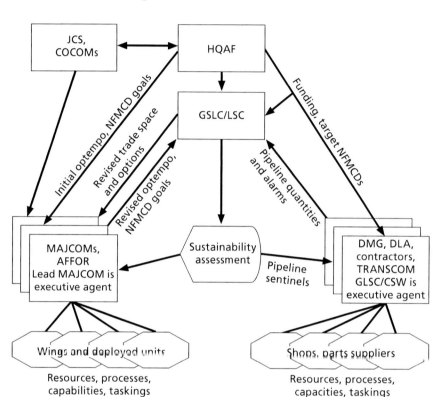

NOTE: The "GLSC/LSC" in the figure represents the Air Force's current transitional
situation, in which implementation of the GLSC is still under way and, therefore,
separate MAJCOM logistics support centers (LSCs) supplement the GLSC.
RAND MG667-4.6

about to falter.[46] Because the measured demand swings may be due to
either changes in the optempo or some variation in the removal rate,
the GLSC would use the information about the recent optempo to
discover the cause. Likewise, the GLSC would focus on those pipeline
MOEIs that the sentinels indicate have exceeded their alarm thresh-

---

[46] Remember that the Adams, Abell, and Isaacson calculations lead to a forecast of future
demands based on forecast future optempo and recent demand patterns. Thus, the CSW
could determine the extent to which a changed optempo might contribute to a future
NFMCD problem by calculating the demands with the original planned optempo.

olds, and they would determine the extent to which demand swings and supply side disruptions like parts shortages or equipment outages contributed to the alarm.

For their part, the demanders (MAJCOMs and AFFOR) may be content monitoring only the NFMCD levels and comparing them to the original planning goals.

Note that we do not advocate that the GLSC take on a full neutral integrator role during execution. Rather, it acts only within the boundaries of the original long-term plans laid out by the Air Force during the POM development process. However, the GLSC may find that some predicted demands do not arise and that the resources allocated to those predicted demands can be used to meet other, unexpected demand surges. Alternatively, supply-side initiatives and workarounds may achieve efficiencies that free up additional resources to meet the changing demands. In that case, the GLSC should have the authority to reallocate resources across the supply-side activities to maintain that originally planned NFMCD level in the POM.

However, when those original resources and total force demands cannot be brought into balance by such technical adjustments within the supply-side activities, the GLSC would play a supporting role to the various parties, developing a new near-term sustainment trade space and alerting the neutral integrator (still the Air Force corporate structure) of the impasse. If direct negotiations between the supply side and the demand side could not resolve the conflict directly, the Air Force corporate structure would then have to decide whether to provide additional resources, or direct the demanders' optempo and NFMCD goals to be reduced.

Adjusting Incentives

As already discussed, adjusting the incentives that may affect all the agencies is beyond the scope of this monograph. However, we would be remiss if we failed to note that publishing goals and measures in an organization invariably causes behavior changes in the organization. Even when those measures are irrelevant to the production process, they induce changes oriented toward them. When the measures are accompanied by perceived or real rewards for improved performance, the reaction is even stronger.

Therefore, we propose no specific incentive changes here other than those that are intrinsic to the measures themselves. In the future, it may be appropriate to investigate how changes in the incentives for all participants from base repair to the depots and contractors might further enhance the performance of this closed-loop planning and execution system.

Periodic Review and Refinement of the COP and Decision Rights

As the technical understanding of the DLR sustainment system changes, as the information (both sharable and latent) available changes, as the decisionmaking capacity evolves, and as the incentives change, one should periodically review the COP design and the decision-rights allocations. One of the most striking changes that occurred during the 20th century was the widespread increase in educational attainment. As just one example, many manufacturing organizations were able to implement statistical process-control techniques in production shops by the end of the century, while typical shops at the beginning of the century had many workers with only rudimentary grade school educations.

Other, subtler, changes may arise if people within the DLR sustainment system agencies obtain more field-level experience. The knowledge they may gain through a COP that gives them greater insights to operations will give them a stronger (albeit latent) understanding of the different kinds of fleet capabilities that the Air Force values and the effort and resources needed to deliver those capabilities.

CHAPTER FIVE

Conclusion

This monograph represents an initial attempt to develop a new process for assuring more effective, more responsive, more efficient materiel sustainment for the U.S. Air Force. We have integrated and described the concepts of common operating pictures, effects-based measures, *schwerpunkt,* decision rights, and the neutral integrator to outline how one might improve the Air Force Materiel Sustainment System. We have illustrated how one might apply those concepts in a specific context, DLR sustainment.

No doubt, there are many ways to improve the specific DLR sustainment system design proposed. After all, it has not been tested in actual operations. We expect that portions of the design may be explored as GLSC's responsibilities and operating processes evolve. Further, we acknowledge that this end-to-end design cannot be implemented as a turnkey change in the operations of DLR support. Rather, it represents a conceptual goal toward which the DLR sustainment system might strive.

At the same time, the document does indicate some enabling mechanisms that could speed the system's evolution toward that goal. If implemented, the Adams, Abell, and Isaacson method of demand forecasting might enable much tighter control and earlier warning of potential surprises. The Hillestad et al. technique for examining the NFMCD and funding trade space across multiple MDSs could make it possible to create operationally relevant, concrete, achievable, and affordable NFMCD goals that reflect Air Force priorities. Finally, sentinels could detect potential problems before the MOEIs increase to levels that would affect those goals.

We anticipate that this design will be gradually refined as it is incorporated within the Air Force DLR sustainment system.

Some Foreseeable Implementation Challenges

This monograph has not dealt with all the potential implementation challenges that a COP-based system might face. Many of these are procedures that, while they make sense in a system without a COP, limit the DLR sustainment system's ability to respond to changing optempos, demand rates, or process changes. Crawford (1988) and Pyles and Shulman (1995) outline the deep uncertainties that have plagued previous attempts to improve DLR sustainment forecasting and control. As described in Chapter Four, the current Air Force use of MICAP-based UND definitions could be replaced by NFMCD-based rules that hold the potential of improving force-wide mission capability. Also, the current financial constraints on repairing DLRs based on two-year-old NSN-specific demand forecasts prevent those same repair resources from being applied to meet actual current demands.

If some variant of this end-to-end, NFMCD-based DLR sustainment control system design were adopted over time, it should be possible to improve the Air Force's control over NFMCD rates, improve the NFMCD balance across fleets to better match operational needs, and ultimately improve those rates force-wide.

That increased flexibility may also require changes to make the DLR sustainment supply chain more flexible and responsive, as well. One reason for adhering to a longer production forecast is that materiel requirements can be predicted with some confidence. If one wants to make the DLR mix more responsive to changing demands, it will be necessary to have the right spare parts available. While the repair process itself can provide some parts responsively—e.g., shop replaceable units on a line replaceable unit—consumable components provided from outside sources are often required, too. To avoid carrying a large inventory of low-turnover repair parts, the Air Force has already begun to investigate ways of improving supply-chain responsiveness in its eLog21 initiative.

Beyond DLRs

We anticipate that the process used in this design may be adaptable to other Air Force materiel sustainment activities such as vehicles, equipment, and ammunition acquisition, storage, and maintenance. Those activities contribute to the six overall Air Force agile combat support objectives identified in Chapter Four in different ways than the activities in support of DLRs. Even so, we anticipate that one could apply the approach outlined in Chapter Three to design common operating pictures to support planning and execution for those activities. Hopefully, these changes will clarify our view that this approach could be used effectively to improve the management of other combat support resources.

How Decisions Occur in Organizations

At some level, organizations exist to make decisions. Almost everything that occurs inside an organization involves a decision of some kind. Asking how decisions occur in organizations comes close to asking how organizations work. That is not our goal in this appendix. Rather, we seek to identify the basic factors that are relevant to making any particular decision and then explain the factors one would consider when assigning decision rights to specific agencies.

In this appendix, we first outline the resources (information and decisionmaking capacity) and influences (incentives) that affect the outcome of any single decision. Then we suggest how one might use those factors to assign decision rights within a large, complex organization.

Information

Any decision typically benefits from information on many different factors. In a large organization, the issue is whether the relevant information is at the right place at the right time. To apply relevant information to a decision, an organization typically has three choices:

- move information from one organizational location to another
- assign the right to make the decision to an organizational location near the most relevant information
- break the decision into smaller decisions, each of which the organization can address at one or more locations near the relevant information.

Some information is latent, that is, it is very difficult to move, at least in time to make the required decision within a meaningful time window. Examples range from the current needs for a part just removed from an aircraft to the deep understanding of the how the complementary operational characteristics of different fleets interact to deliver a military capability. On one hand, information may be latent because it requires direct observation and immediate action. On the other, it may require extensive education or experience that is not widely available throughout the organization.

In many organizations, advances in information technology have made it possible to move much more information more readily and rapidly. In addition, increased access to education over the past century has also made some previously scarce deep information much more widely available, making it less necessary to move.

These technological and educational shifts have increased the range of options available to assign decision rights to different organizational elements. In many cases, organizations have exploited those changes to decentralize the rights to make various decisions. Yet, those same organizations have maintained a considerable degree of centralization for other decisions. The reasons for these different organizational design choices lie in the incentives and the decisionmaking capacity available.

Information Processing Capacity and Incentives

A decision settles a question. Within an organization, a decision in effect chooses one course of action and sets the organization on that course. Thus, every decision in an organization occurs in a specific place at a specific time. A decision may be the product of many discussions and interactions, among many organizations and individuals within organizations, over a long period of time. But somewhere, at some point in time, settlement occurs, changing or clarifying the direction of future actions in the organization.

Given any particular decision of the kind listed above, who should have authority to make or influence the decision in an organization? Presumably the person or activity in that organization most likely to make a decision that advances the goals of the organization should have the authority to make the decision. If other individuals or activities can provide input to making the decision that is likely to promote the organization's goals, they should have some authority to provide that input.

A decisionmaker is more likely to benefit the organization if the decisionmaker

- knows more about what sets of outcomes of a decision are most *desirable* from the organization's perspective, or most likely to benefit the organization
- knows more about what sets of outcomes alternative decisions can achieve, or what is *feasible*
- has an *incentive* to choose the set of outcomes that benefits the organization most.

Put another way, the more relevant *information* the decisionmaker has and the more *incentivized* the decisionmaker is to use that information to advance the organization's goals, the more likely that decisionmaker is to benefit the organization.

The quality of the information available to a decisionmaking process at any organizational location depends on two things:

- How good is the information received from relevant sources— monitors of strategic priorities, the military situation, the general environment, and the status of resources and processes?
- How good is the local capability to integrate information from all these sources in a way that creates an image relevant to the organization's strategic goals? This capability depends on the level of decisionmaking skills and the methods and tools available at the location—that is, the decisionmaking capacity.

The information that a decisionmaker can actually use is information that the decisionmaker and his or her staff have processed and absorbed locally in a way that prepared it properly for application to the decision at hand.

Assigning Decision Rights

Organizations exist to create synergies among many different kinds of specialists. Specialist communities within an organization nurture and promote different perspectives on the relative importance of various strategic organizational goals and information on how firm or binding the different constraints may be.

Effective incentives to promote the organization's goals also differ from location to location. The mechanisms used to motivate individuals often reflect local values, and individuals typically retain some autonomy to pursue their own interests, even if they depart from the organization's interests. The degree of this autonomy often differs at various locations in the organization.

Who, then, should the organization give the authority to make specific decisions? The answer depends on which information about strategic values and specific constraints is most relevant to a specific decision. The better the relevant information and decisionmaking capacity is at a particular location, the more desirable it is to place a decision at that location. The answer also depends on how good the incentives are at different locations in the organization. If the information relevant to a location is perfect, incentives that do not constrain pursuit of local or individual interests there can make that location an unattractive place to make a decision. Similarly, even if the incentives at a location perfectly align a decisionmaker's values with the organization's values, that location may be unattractive if information relevant to the decision is missing.

References

Ackerman, Robert K., "Operation Enduring Freedom Redefines Warfare," *Signal Magazine*, September 2002, p. 3.

Adams, John L., John B. Abell, and Karen Isaacson, *Modeling and Forecasting the Demand for Aircraft Recoverable Spare Parts*, Santa Monica, Calif.: RAND Corporation, R-4211-AF/OSD, 1993. As of November 9, 2007:
http://www.rand.org/pubs/reports/R4211/

Alberts, David S., and Richard E. Hayes, *Power to the Edge: Command Control in the Information Age*, Washington, D.C.: Command and Control Research Program, Office of the Assistant Secretary of Defense, 2003.

Bearing Point, *U.S. Air Force Logistics Enterprise Architecture Concept of Operations v 1.6.1*, Washington, D.C.: Bearing Point, August 1, 2005.

Bester, Helmut, "Externalities and the Allocation of Decision Rights in the Theory of the Firm," CEPR Discussion Paper No. 3276, March 2002.

Boyd, John, "Patterns of Conflict," briefing presented to U.S. Air War College, Maxwell Air Force Base, Ala.: Air University Press, 1986.

Clausewitz, Carl von, *On War*, ed. and tr. Michael Howard and Peter Paret, Princeton, N.J.: Princeton University Press, [1832] 1984.

Cone, BG Robert W. (U.S. Army), "Briefing on Joint Lessons Learned from Operation Iraqi Freedom," news transcript, October 2, 2003. As of November 9, 2007:
http://www.defenselink.mil/transcripts/transcript.aspx?transcriptid=3531

Crawford, Gordon, *Variability in the Demand for Aircraft Spare Parts: Its Magnitude and Implications*, Santa Monica, Calif.: RAND Corporation, R-3318-AF, 1988. As of November 9, 2007:
http://www.rand.org/pubs/reports/R3318/

Dail, MGen Robert T., testimony before the House Armed Services Committee Subcommittee on Readiness, March 30, 2004.

Davis, Paul K., *Effects-Based Operations (EBO): A Grand Challenge for the Operations Community*, Santa Monica, Calif.: RAND Corporation, MR-1477-USJFCOM/AF, 2001. As of November 9, 2007:
http://www.rand.org/pubs/monograph_reports/MR1477/

Deptula, BGen David, *Effects-Based Operations: Changes in the Nature of Warfare*, Arlington, Va.: Aerospace Education Foundation, 2001.

Fischbein, Martin, and Icek Ajzen, *Belief, Attitude, Intention and Behavior: An Introduction to Theory and Research*, Reading, Mass.: Addison-Wesley, 1975.

Gabreski, Brig Gen Terry L., *Chief of Staff, United States Air Force Logistics Review*, Washington, D.C.: Headquarters, Air Force/ILM, June 2000.

Gallagher, Maj Gen Patrick, "AF POM Development: AFSO-21 Rapid Improvement Event," briefing, Headquarters, Air Force/A8, Washington, D.C., March 2007.

Garvin, David A., and Michael A. Roberto, "What You Don't Know About Making Decisions," *Harvard Business Review*, Vol. 79, No. 8, September 2001, pp. 108–116.

Glasser, Perry, "Armed with Intelligence," *CIO Magazine*, August 1997, pp. 104–110.

Headquarters, Air Force Materiel Command, *Standard Practice for System Safety*, MIL-STD882D, Ohio: Wright-Patterson Air Force Base, February 10, 2000.

Headquarters, Air Force Materiel Command, A4Y, *Execution and Prioritization Repair Support System (EXPRESS)*, Instruction 23-120, Ohio: Wright-Patterson Air Force Base, May 24, 2006.

Headquarters, Air Force Materiel Command, A8R, *Depot Maintenance Accounting and Production System—Financial Policy and Procedures for Organic Depot Maintenance*, Instruction 65-101, Ohio: Wright-Patterson Air Force Base, March 28, 2006.

Headquarters, Air Force Materiel Command, FMRF, "Future Financials (FF) PMO Update," briefing, Ohio: Wright-Patterson Air Force Base, May 15, 2006.

Headquarters, U.S. Air Force, *Air Force Roadmap 2006–2025*, June 2006. As of November 14, 2007:
http://www.af.mil/shared/media/document/AFD-060713-002.pdf

Hillestad, Richard, *Dyna-METRIC: Dynamic Multi-Echelon Technique for Recoverable Item Control*, Santa Monica, Calif.: RAND Corporation, R-2785-AF, 1982. As of November 9, 2007:
http://www.rand.org/pubs/reports/R2785/

Hillestad, Richard, Robert Kerchner, Louise W. Miller, Adam C. Resnick, and Hyman L. Shulman, *The Closed-Loop Planning System for Weapon System Readiness*, Santa Monica, Calif.: RAND Corporation, MG-434-AF, 2006. As of November 9, 2007:
http://www.rand.org/pubs/monographs/MG434/

Isaacson, Karen E., Patricia Boren, Christopher L. Tsai, and Raymond A. Pyles, *Dyna-METRIC Version 4: Modeling Worldwide Logistics Support of Aircraft Components*, Santa Monica, Calif.: RAND Corporation, R-3389-AF, 1988. As of November 9, 2007:
http://www.rand.org/pubs/reports/R3389/

Isaacson, Karen E., and Patricia Boren, *Dyna-METRIC Version 6: An Advanced Capability Assessment Model*, Santa Monica, Calif.: RAND Corporation, R-4214-AF, 1993. As of November 9, 2007:
http://www.rand.org/pubs/reports/R4214/

Jensen, Michael C., "Put the Right Decisions in the Right Hands," *Harvard Management Update*, Vol. 10, No. 5, May 2005.

Keating, Edward G., *Government Contracting Options: A Model and Application*, Santa Monica, Calif.: RAND Corporation, MR-693-AF, 1996. As of November 9, 2007:
http://www.rand.org/pubs/monograph_reports/MR693/

King, Lieutenant General James C. (U.S. Army), "Keynote Address to the ASPRS 2000 DC Annual Conference," *Photogrammetric Engineering and Remote Sensing*, Vol. 66, No. 9, September 2000, pp. 1043–1047.

Kirby, William, "Centralized Asset Management Program Overview," briefing, Ohio: Wright-Patterson Air Force Base, February 27, 2007.

Leftwich, James A., Robert S. Tripp, Amanda B. Geller, Patrick Mills, Tom LaTourrette, Charles Robert Roll, Jr., Cauley Von Hoffman, and David Johansen, *Supporting Expeditionary Aerospace Forces: An Operational Architecture for Combat Support Execution Planning and Control*, Santa Monica, Calif.: RAND Corporation, MR-1536-AF, 2003. As of November 9, 2007:
http://www.rand.org/pubs/monograph_reports/MR1536/

Loredo, Elvira N., Raymond A. Pyles, and Don Snyder, *Programmed Depot Maintenance Capacity Analysis Tool: Workloads, Capacity, and Availability*, Santa Monica, Calif.: RAND Corporation, MG-519-AF, 2007. As of November 9, 2007:
http://www.rand.org/pubs/monographs/MG519/

Lynch, Kristin F., John G. Drew, David George, Robert S. Tripp, Charles Robert Roll, Jr., and James A. Leftwich, *The Air Force Chief of Staff Logistics Review: Improving Wing-Level Logistics*, Santa Monica, Calif.: RAND Corporation, MG-190-AF, 2005. As of November 9, 2007:
http://www.rand.org/pubs/monographs/MG190/

McCammant, LtCol Chris, *Background Paper on Centralized Asset Management*, Ohio: Wright-Patterson Air Force Base, October 23, 2006.

Miller, Louise W., Richard E. Stanton, and Gordon Crawford, *Dyna-Sim: A Nonstationary Queuing Simulation with Application to the Automated Test Equipment Problem*, Santa Monica Calif.: RAND Corporation, N-2087-AF, 1984. As of November 9, 2007:
http://www.rand.org/pubs/notes/N2087/

Mills, Patrick, Ken Evers, Donna Kinlin, and Robert S. Tripp, *Supporting Air and Space Expeditionary Forces: Expanded Operational Architecture for Combat Support Planning and Execution Control*, Santa Monica, Calif.: RAND Corporation, MG-316-AF, 2006. As of November 9, 2007:
http://www.rand.org/pubs/monographs/MG316/

Morrill, MGen Art, "Lean Thinking 101," briefing, Ohio: Wright-Patterson Air Force Base, 2006.

Myers, General Richard B. (USAF), "A Word From the Chairman," *Joint Forces Quarterly*, No. 33, Winter 2002–2003, pp. 1, 4–8.

Pyles, Raymond A., and Robert S. Tripp, *Measuring and Managing Readiness: The Concept and Design of the Combat Support Capability Management System,* Santa Monica, Calif.: RAND Corporation, N-1840, 1982. As of November 9, 2007:
http://www.rand.org/pubs/notes/N1840/

Pyles, Raymond A., *The Dyna-METRIC Readiness Assessment Model: Motivation, Capabilities, and Use*, Santa Monica, Calif.: RAND Corporation, R-2886, 1984. As of November 9, 2007:
http://www.rand.org/pubs/reports/R2886/

Pyles, Raymond A., and Hyman L. Shulman, *United States Air Force Fighter Support in Operation Desert Storm*, Santa Monica, Calif.: RAND Corporation, MR-468-AF, 1995. As of November 9, 2007:
http://www.rand.org/pubs/monograph_reports/MR468/

Sherbrooke, Craig C., *METRIC: A Multi-Echelon Technique for Recoverable Item Control*, Santa Monica, Calif.: RAND Corporation, RM-5078-PR, 1966. As of November 9, 2007:
http://www.rand.org/pubs/research_memoranda/RM5078/

———, *Evaluation of Demand Prediction Techniques*, Bethesda, Md.: Logistics Management Institute, Report AF601R1, March 1987.

Slay, F. Michael, and Craig C. Sherbrooke, *The Nature of the Aircraft Component Failure Process: A Working Note*, Bethesda, Md.: Logistics Management Institute, Report AF701R1, February 1988.

Slay, F. Michael, *Demand Forecasting*, McLean, Va.: Logistics Management Institute, Briefing Book AF401LN2, May 1995.

Tripp, Robert S., *Supporting Expeditionary Aerospace Forces: An Integrated Strategic Agile Combat Support Planning Framework*, Santa Monica, Calif.: RAND Corporation, MR-1056-AF, 1999. As of November 9, 2007:
http://www.rand.org/pubs/monograph_reports/MR1056/

Tripp, Robert S., Kristin F. Lynch, John G. Drew, and Edward W. Chan, *Supporting Air and Space Expeditionary Forces: Lessons from Operation Enduring Freedom*, Santa Monica, Calif.: RAND Corporation, MR-1819-AF, 2004. As of November 9, 2007:
http://www.rand.org/pubs/monograph_reports/MR1819/

Tripp, Robert S., Kristin F. Lynch, Ronald G. McGarvey, Don Snyder, Raymond A. Pyles, William A. Williams, and Charles Robert Roll, Jr., *Strategic Analysis of Air National Guard Combat Support and Reachback Functions*, Santa Monica, Calif.: RAND Corporation, MG-375-AF, 2006a. As of November 9, 2007:
http://www.rand.org/pubs/monographs/MG375/

Tripp, Robert S., Kristin F. Lynch, Charles Robert Roll, Jr., John G. Drew, and Patrick Mills *A Framework for Enhancing Airlift Planning and Execution Capabilities Within the Joint Expeditionary Movement System*, Santa Monica, Calif.: MG-377-AF, 2006b. As of November 9, 2007:
http://www.rand.org/pubs/monographs/MG377/

U.S. Air Force, *Global Logistics Supply Center*, Air Force Internal Working Paper, July 2006. Not available to the general public.

———, *Implementation of the Air Force Chief of Staff Direction to Establish an Air Force Component Organization*, Air Force Internal Working Paper, AF PAD 06-09, November 7, 2006. Not available to the general public.

U.S. Air Force Logistics Transformation Team, *Chief of Staff Logistics Review (CLR) Centralized Intermediate Repair Facilities (CIRF) Test Plan 1 September 2001–1 March 2002*, Washington, D.C.: Deputy Chief of Staff, Installation and Logistics, April 2001.

———, *Determining Manpower Requirements*, Washington, D.C.: U.S. Air Force, Air Force Instruction 38-201, 1999.

U.S. Department of Defense, *Quadrennial Defense Review Report 2006*, Washington, D.C.: U.S. Department of Defense, 2006.

Wainfan, Lynne, and Paul K. Davis, *Challenges in Virtual Communication: Videoconferencing, Audioconferencing, and Computer-Mediated Communications*, Santa Monica, Calif.: RAND Corporation, MG-273, 2004. As of November 9, 2007:
http://www.rand.org/pubs/monographs/MG273/

Warden, Col. John A. III, *The Air Campaign: Planning for Combat*, Washington D.C.: National Defense University Press, 1988.

Zettler, Lt Gen Michael, "A View from the Top," letter, Headquarters, U.S. Air Force, November 24, 2003.